International Conference on Wind Propulsion 2021

Online
15-16 September 2021

ISBN: 978-1-7138-8311-1

Printed from e-media with permission by:

Curran Associates, Inc.
57 Morehouse Lane
Red Hook, NY 12571

Printed with permission by Curran Associates, Inc. (2024)

The Royal Institution of Naval Architects
8-9 Northumberland Street
London, WC2N 5DA
United Kingdom

Phone: 020 7235 4622
Fax: 020 7259 5912

publications@rina.org.uk

Additional copies of this publication are available from:

Curran Associates, Inc.
57 Morehouse Lane
Red Hook, NY 12571 USA
Phone: 845-758-0400
Fax: 845-758-2633
Email: curran@proceedings.com
Web: www.proceedings.com

International Conference

Wind Propulsion 2021

15-16 September 2021

PROCEEDINGS

Sponsored by:

THE ROYAL INSTITUTION OF NAVAL ARCHITECTS
8-9 Northumberland Street, London, WC2N 5DA, Telephone: +44 (0)20 7235 4622

The Royal Institution of
Naval Architects

INTERNATIONAL CONFERENCE

Wind Propulsion 2021

15-16 September 2021, Online Conference

Papers published in these conference Proceedings are peer reviewed.

THE ROYAL INSTITUTION OF NAVAL ARCHITECTS
8-9 Northumberland Street
London WC2N 5DA

Telephone: 020 7235 4622
Fax: 020 7259 5912

CONTENTS

TRANSITION WIND TECHNOLOGIES IN SHIPPING TO 2050: FACTORS AND CHALLENGES FOR A SUSTAINABILITY TRANSITION

J Köhler and **E Dönitz Fraunhofer,** ISI, Germany
F Schätter, Hochschule Pforzheim, Germany

SUMMARY

This paper presents a scenario for a transition to low carbon shipping. It combined a structured scenario process and simulations with the MATISSE-SHIP model to develop consistent qualitative and quantiative scenarios, agreed with sector experts. The IMO decides a binding target to reduce GHG emissions by 90 % by 2050. The use of LCF technologies for shipping, such as PtL/PtG, H_2 fuel cells, and wind technologies, becomes the state of the art. Rapid action is required by the shipping sector. If society is willing to prioritise reduction of climate change, new operational patterns in shipping are possible. This behavioural change needs to be supported by strong climate policy to force the rapid development and adoption of wind and other low carbon technologies.

NOMENCLATURE

ABM	agent-based model
GHG	greenhouse gas
HFO	Heavy fuel oil
IMO	International Maritime Organisation
LCF	Low carbon fuels
LfL	Low Flashpoint Liquid fuel
LNG	Liquid natural Gas Dual Fuel motor
MDO	Marine diesel oil
MGO	Marine gas oil
PtL/PtG	power-to-liquid/gas synthetic fuels
UNFCCC	United Nations Framework Convention on Climate Change

1. INTRODUCTION

Shipping is a major contributor to global greenhouse gas (GHG) emissions and the rate of growth assumed in the published literature is high, such that emissions from shipping are expected to increase in absolute terms into the future, possibly by 50-250 % [1]. There are few medium to long term studies that provide scenarios of major reductions in GHG emissions and these are also limited in their ambition. For example, [2] only assumes a potential for slow steaming of 20 % energy saving by 2050. [3] only has a time horizon until 2030, such that fuels such as LNG are seen as the main alternative to MDO/MGO or low sulphur HFO. These studies do not fully exploit the potential of more radical long term changes such as large scale H_2, wind assistance/propulsion in suitable trades or combinations of technologies such as H_2 and batteries or PtL/PtG technologies.

This paper uses the AHOY Scenarios [4] for scenarios for shipping to 2050. It used a structured scenario process to identify the critical factors that determine scenarios of GHG mitigation and estimated the potential for shipping to contribute to GHG mitigation globally, by developing scenarios of the development and diffusion of technologies to reduce GHGs from shipping. The use of the new MATISSE-SHIP computer simulation model [5,6] used the assumptions to quantify scenarios of technology development in shipping, taking into account the assessments of industry experts and including a wide range of power/propulsion technologies.

The factors determining the development of propulsion technologies in shipping are described and the combination that led to a transition where wind plays a major role presented. The scenario results are shown and challenges for the development of wind technologies discussed.

2. FACTORS FOR A TRANSITION TO WIND

General factors that determine the contexts for the scenarios have been structured using the STEEPL (social, technological, environmental, economic, policy and legal) approach. Thirteen key factors were identified, shown in table 1. Key factors such as "environmental awareness and regulation", "global economic development" or "climate change" are the critical elements in the whole system: these elements influence the other elements most strongly and are also influenced by them most strongly at the same time. Moreover, key factors like "naval architecture" or "electricity storage technologies" are buffering elements, i.e. they influence the remaining elements weakest and are influenced by them weakest. Finally, the key factor "digital technologies" influences almost all other factors, but in the same time the factor is independent from other developments.

3. A SCENARIO OF TECHNOLOGICAL AND BEHAVIOURAL TRANSITION

The AHOY 'Y' scenario projects a world of both technological and behavioural change. Society experiences a fundamental change of mind-set toward sustainable lifestyles and increased environmental awareness. Extrene weather events and resulting waves of migration cause a social acceptance that climate protection is of crucial importance. New social concepts and behaviors, e.g. strengthened focus on work life balance or job sharing, as well as sustainable consumption are established in society. This shift of paradigm also means an end of conventional global economic growth. The focus is shifted to qualitative growth in sustainable goods and services. The consequence of these developments is a strong global response to climate change on both international (e.g. UNFCCC) and national levels and the mandatory implementation and achievement of the Paris Agreement and the 1.5°C target.

The resulting wave of investment in low carbon technologies offsets the reduction in demand in the conventional economy. The IMO reacts and decides a binding target to reduce GHG emissions by 90 % by 2050, which requires a pathway of 50% reduction by 2035. Governments fund the development of green technologies. The marine industry is driven by new materials, such as non-steel ships, ship constructions that are recyclable or made of light carbon fiber materials. The requirement for rapid decarbonization is met through large-scale retrofitting for low carbon technologies. The demand to stop the construction of ships with fossil propulsion systems prevails; lLegal frameworks globally prohibit the usage of fossil fuel in the transport sector from 2030.

3.1 OPPORTUNITIES FOR GROWTH

As industries, logistics, and cruise passengers set their priority on clean shipping, the shipping sector gets under very strong pressure. To ensure energy supplies, politics sets the legal framework conditions to build an adequate LCF infrastructure on a national and continental level. By accelerating the growth of electricity supply from renewable energies, rapid development of PtL/PtG in shipping is made possible. Retrofit markets for both PtL/PtG and wind technologies are an important part of rapid decarbonization.

These measures are successful in achieving a very rapid transition away from fossil fuels in transport. The use of LCF technologies for shipping, such as PtL/PtG, H_2 fuel cells, and wind technologies, becomes the state of the art. Biofuels do not take off, because of the very strong demand from the aviation sector and social concerns about land use for the production of biofuels. Digital technologies become an important factor for shipping: Sensors on board allow condition monitoring and navigation, autonomous collision avoidance, course control given wind and sea state and technical support for crews on-board. There are also remotely operated ships where no crew is required with ship management centers reducing the requirement for manned operation. Digitalization also implies that ship systems must be hardened against digital threats and made highly redundant.

3.2 MATISSE-SHIP SCENARIO

The scenario (see figure 1) shows the fastest decline in fossil fuels of all the AHOY scenarios. strong regulatory requirements for a reduction in both NO_x /SO_x and CO_2 are complemented by a recognition that rapid, far-reaching changes in shipping are necessary. The general cultural shift toward sufficiency and a high priority on climate change mitigation mean that lower operating speeds are required, to some extent even in cruise/ferry, container and service vessels. Bulker/tanker ship-owners face increasing costs for conventional ships and fuels.

LNG peaks in 2030 at a low level due to the fossil fuel ban in 2030. LCF technologies will dominate the market with 90% by 2050. Depending on ship types, in 2050 PTl/PtG fuels will almost have the same market share as Wind/H2 hybrids, where fuel cells provide power for auxiliary systems, emergency and manoeuvring.

4. CHALLENGES AND IMPLICATIONS FOR SHIPPING

Based on experts' judgements and discussions, challenges arising across the scenarios and their implications for the shipping sector were identified. Main challenges relevant to all scenarios are: piracy and cyber-attacks, central banking interest rates, the global willingness to pay for change, the issue of a lacking business cases for the adoption of low carbon propulsion, the competition between different sectors and the problem of ineffective changes of CO_2 footprints. The challenges and implications for the transition to low carbon shipping and wind technologies are shown in table 2.

5. CONCLUSIONS

The combination of structured qualitative scenario analysis and simulation modelling provides a consistent scenario with a clear statement of the assumptions for the scenario analysis. The scenario was agreed with the stakeholder process for the AHOY scenarios [4]. The assessment of challenges shows that there are number of challenges for the shipping industry and society to address, if such radical changes are to happen so quickly. The main conclusions are:

i. **Rapid action is required:** IMO targets are reached via a fossil fuel ban by 2030. The Paris Agreement is enforced around the globe.
ii. **Technology follows regulation:** If politics sets the legal framework to build an adequate LCF infrastructure on a national and global level, technology follows. This enables the rapid development of PtL/PtG in shipping. The demand for rapid decarbonization is met by the widespread introduction of low-carbon technologies in both newbuildings and retrofits.
iii. **Social attitudes matter:** Global regulation and successful decarbonization can be driven forward by a change in awareness in society.

iv. **Attitude change can cause changes in logistics:** The acceptance of lower operational speeds allows wind and low carbon technologies with lower energy storage densities to grow and develop a major market share.
v. **For the shipping industry:** if customers set their priority on clean shipping, the industry must be prepared to act very quickly.

6. ACKNOWLEDGEMENTS

Funding for this research by MAN Energy Solutions AG is gratefully acknowledged.

7. REFERENCES

[1] IMO, 'Third IMO GHG Study 2014'. MEPC 67/INF.3. IMO. London, 2014 Online verfügbar unter http://www.imo.org/en/OurWork/Environment/PollutionPrevention/AirPollution/Pages/Greenhouse-Gas-Studies-2014.aspx.

[2] DNV-GL, 'Low Carbon Shipping towards 2050', DNV-GL 2017

[3] Lloyds; UCL 'Global Marine Fuel Trends 2030', Lloyds Register Marine and University College London, 2014.

[4] MAN Energy AG, *#AHOY2050 Exploring scenarios for the maritime industry's pathway to a greener future*, MAN Energy AG, 2020.

[5] Köhler J., 'Modelling the multi-level perspective: The MATISSE agent-based model', In Enayat Moallemi, F. de Haan (Eds.): Modelling Transitions – Virtues, Vices, Visions of the future: Routledge: Oxon; New York pp. 77-101, 2020.

[6] Köhler J., 'Zero carbon propulsion in shipping - scenarios for the development of hydrogen and wind technologies with the MATISSE-SHIP model, International Shipbuilding Progress, 67(1), 79-95, 2020.

8. AUTHORS BIOGRAPHY

Dr. Jonathan Köhler holds the current position of senior scientist at Fraunhofer ISI. His previous experience includes the position of Constructor, Royal Corps of Naval Architects and manager, Integrated Assessment Model Development for the Tyndall Centre.

Dr. Ewa Dönitz is a senior researcher at Fraunhofer ISI.. Her research focusses on the whole range of Foresight methods in particular scenarios, visions and roadmaps and its application for underpinning strategy building and innovation processes.

Dr. Frank Schätter is Professor of Supply Chain Processes Management at the Faculty of Business and Law at Pforzheim University. He has advised companies of different industries and associations in questions of strategic network planning and future-oriented strategy development.

Table 1 Initial practices of technologies

STEEPL Factor	**Future assumptions for scenario of technological and behavioural transition**
Environmental Awareness	Sustainable lifestyle through cultural change
Global economic developments	End of conventional global growth
Climate change	Strong climate policy at the UNFCCC (2°C Paris target attained)
Oil price	Oil price not relevant, as fossil fuels prohibited in the transport sector
Environmental regulation	To meet the 2°C objective, reduce shipping GHGs by 80% by 2050
Political crises	Globalisation with tensions (staus quo)
Digital technologies	Diffusion of digital technologies imply threats, slow development
Low carbon technologies	Wind and synthetic fuels become the main alternative
Energy infrastructure	Policy driven development of renewable energy infrastructure
Electricity storage capacity	Limtied application in short sea shipping
Requirements for logistics network	Increase of supply chain complexity
Naval architecture	Push through digitalisation
Autonomous shipping and regulation	Ships with digital support

Table **2** Challenges and implications for shipping

Challenge	Implications for shipping
Piracy and cyber-attack as a new form of piracy	• The challenge becomes particularly crucial when autonomous shipping becomes state of the art • Developing anticipatory scenarios for threats to digital infrastructure involve the consideration of a global attack surface • Policy must be continuously updated to address future security risks
Central banking interest rate	• When interest rates increase it is more beneficial for ship-owners to run their ships with a higher speed in order to reduce the extra cost of interest payments on financing for the goods in transit on ships • Trade-off between saving fuel costs by sailing slower and the extra costs of crew, lower transport capacity per ship and other costs • If fossil fuel prices are high and the alternative fuels are also expensive, the "A" and "O" scenarios will have a balance more in favor of slower steaming or sail
Technology development until 2050	• To achieve a high amount of LCF-operated ships in 2050, the deadline for new technology installation should be actually 2030 • This is because of the long life of ships as well as extensive retrofitting
Global willingness to pay for change	• Decarbonization of shipping will be supported if consumers are willing to pay more for green products • Tension between what is needed and what people want • Just when oil taxes are implemented globally (UNFCCC), a level playing field is guaranteed
Total lifecycle of propulsion system	• Shipping must consider the overall environmental impact of biofuels vs. PtL/PtG or other alternative; consider CO_2 footprint of fuels and engines of the whole value chain & life cycle • Not relevant for scenario "O" as there is no strong climate policy globally; the consideration would only be important if the EU adopts strong climate policy.
Business case for adoption of low carbon propulsion (losing market opportunities due to international environment friendly standard setting)	• Depends on the long-term financial health of the shipping industry and development of global freight transport demand • Business case depends on environmental policy • Key questions: what are the financial incentives for companies? Is there a change of business models from fossil fuels to LCF? • Development costs for alternative concepts must be considered (who pays tech)
Competition between different sectors	• Competition between shipping and other sectors, especially for biofuels (food vs. fuel; biofuels vs reforestation) • Also competition for possible limited renewable power for PtL/PtG for shipping vs. other sectors (e.g. PtL/PtG is used for chemical processes)
Political discussion on CO2 footprint	• There are a range of views in the society about the importance and feasibility of effective climate policy • Difficulty that society and politics keep debate alive, but current developments in extreme climate events and the new popular movements suggest that the debate is becoming more intense
No change of CO2 footprint	• Number of ships and cargo volume increases when economy grows, so reduction in 'absolute' levels of emissions is more difficult if technology is not adopted • When there is more growth, there are more financial resources for innovation and more new ships that can be built with low carbon propulsion • Also relevant for "Y" scenario because it assumes lower (conventional economic) growth
Environmentally aware consumers (carbon efficient)	• There might be different consumers markets with varying requirements for shipping as a logistics provider for transport, including requirements for better environmental performance in the supply chain including shipping • The transportation sector is seen as a major contributor to global emissions; companies operating in this sector must continue to be sensitive to the shift in public opinion towards companies classified as "polluting"
Climate impacts become more severe	• Shipping has to deal with climate impacts, e.g. rise of sea level, storms, harbor infrastructure
Application of advanced materials	• Potential to use new materials for ships with long lifecycle; existing ships retrofit • For lower displacement and higher performance systems, including lower maintenance requirements which will reduce the workload of crews and/or make implementation of remote controlled or autonomous ships easier
High power requirements (e.g. Cruise with 60 MW+) with alternatives	• Biofuels or PtL/PtG needed, wind or fuel cells will require very radical solutions to provide this level of power for a whole voyage
Battery storage capacity / power density	• Batteries are relatively heavy compared to liquid fuels, so with the current technology, they can only be used for short distances or as auxiliaries • Technical challenge

Figure 1: MATISSE-SHIP scenario Behavioural and technological change

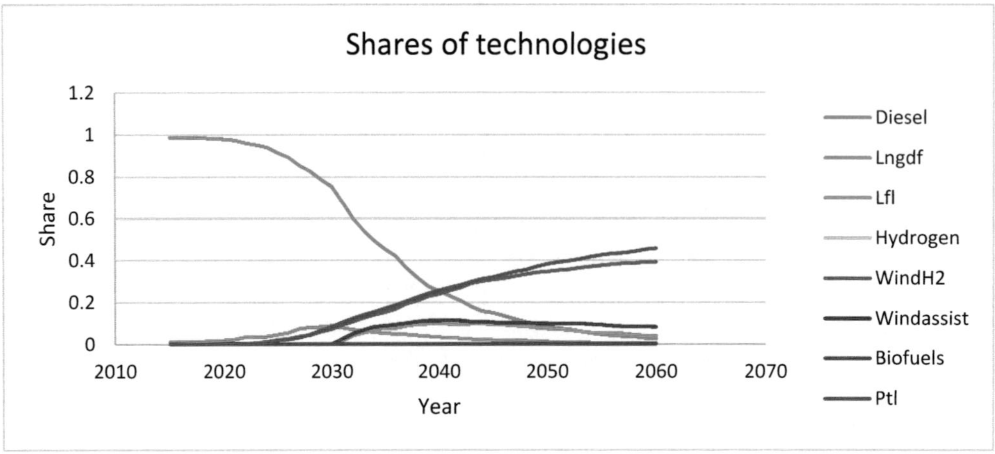

HORSES FOR COURSES: HOW TO SELECT THE "RIGHT" WIND PROPULSION SYSTEM AND HOW TO MAKE THE BUSINESS CASE

F C Gerhardt, S Werner, A Hörteborn, O Lundbäck, J Nisbet, T Olsson, SSPA Sweden AB, Sweden

SUMMARY

Wind propulsion systems (WPS) are major investments and the decision to install them requires careful consideration of many complex questions. In this paper we present a systematic, scientific methodology to assess the benefits and drawbacks of such systems at the early concept stage of a vessel. The purpose is to provide guidance for shipowners and operators and help them make informed decisions. The proposed method was developed into a Software tool called 'SEAMAN Winds' and has been correlated to full scale results. The program draws on our large database of model tests, and CFD of hulls and wind propulsion technologies. It uses the intended trading routes of the vessel as an important input, typical output data are:

a) performance values (ship speed, power requirements etc.)

b) environmental parameters (CO_2 avoided, EEDI and EEXI reduction, carbon intensity indicator)

c) financial metrics (bunker savings, payback time for installation of WPS)

Potential applications of the method include making the business case for one particular WPS or investigating in how far certain systems are more suited for a specific route than others.

1. INTRODUCTION

1.1 BACKGROUND

Due to the many advantages of Diesel propulsion and the availability of cheap oil interest in commercial sailing vessels has been minimal for most of the 20th century. Now, at the beginning of the 21st century the climate crisis has changed this.

In 2018 international shipping accounted for around 2.89% of globally produced CO_2 emissions. Forecasts about a future increase range from 90% to 130% by 2050 compared to the baseline year of 2008. Such disquieting projections have placed the topic of decarbonising shipping on the forefront of the global policy making agenda. The International Maritime Organizations (IMO) reached a milestone agreement in April 2018. In alignment with the emission reduction goals set out in United Nation's 2015 Paris Agreement, IMO agreed to cut greenhouse gas emissions by at least 50% by 2050 and CO_2 emissions per transport work by at least 40% by 2030 and 70% by 2050 compared to the 2008 level. To achieve such ambitious goals the international community is currently examining various measures and actions to enhance the energy efficiency of the maritime transport sector [1]

Bouman et al. [2] have recently studied the CO_2 reduction potential of the following main categories of improvement measures:

- Hull design
- Power & propulsion system
- Alternative fuels
- Alternative energy sources
- Operational measures

Figure 1 summarises their findings. As can be seen, wind propulsion systems (WPS) are among the more promising technologies to reduce CO_2 emissions.

Several WPS-concepts are currently being developed and re-discovered. Such modern technologies have little in common with the canvas sails of old and range from "Flettner-rotors" over kites and suction sails to rigid sails that resemble vertical aircraft wings [1].

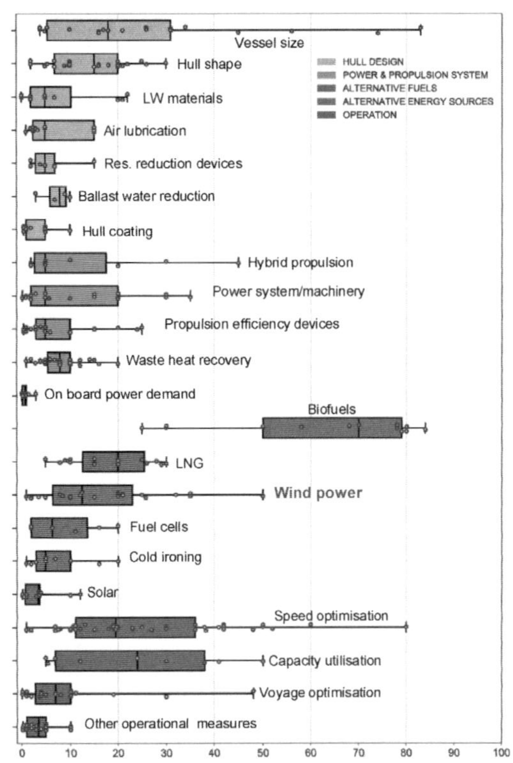

Figure 1: CO_2 reduction potential of various measures, from [2], reprinted with permission.

1.2 MODERN WIND PROPULSION SYSTEMS

Wind propulsions systems can be grouped into seven main categories, [3]:

1. Soft Sails: Traditional cloth & modern adaptions
2. Hard Sails: Wingsails w & w/o flaps
3. Rotor Sails: Based on the "Magnus"-effect
4. Suction Sails: Non-rotating wing with boundary layer suction to increase maximum lift values.
5. Kite Sails: Flown off the bow to assist propulsion
6. Turbines: Wind turbines to generate electrical energy.
7. Hull Form: Redesign of ship's hulls to capture the wind and generate thrust.

All of these systems have their specific strengths and weaknesses which need to be considered when selecting a WPS for a specific case. For the more common systems (1.-5. in above list) Table 3 at the end of the paper summarises just some of the complex questions involved in choosing the "optimal" WPS.

To help shipowners and other stakeholders make informed decisions this paper illustrates a systematic, scientific methodology to compare and assess different WPS at the early concept stage of a vessel. The methodology takes technical performance questions, the trading routes of the vessel and financial & risks aspects into account and is intended to help making the "business case".

3. DECISION SUPPORT METHODOLOGY

An introduction to the proposed WPS assessment methodology and the resulting software tool, called SEAMAN **Wi**nd **in**vestment **d**ecision **s**upport (*SEAMAN Winds*) follows below.
After a description of the background and modelling assumptions, *SEAMAN Winds* is used to study the case of 150-metre long RoPax ferry.

3.1 OVERVIEW

Figure 3 illustrates the methodology behind *SEAMAN Winds*. The program consists of four modules that look at ship performance, voyage & routing, financial & environmental, and risk aspects respectively.
The approach is similar to methods developed elsewhere [4, 5, 6] but differs in the following respects:

- Hull and propeller modelling is based on SSPA's large database of experimental and CFD results, as a complement to the usual empirical methods.
- The methodology also includes risk aspects.
- It has been developed in dialog with end-users from the industry and has a commercial rather than an academic or research focus.
- It has successfully been used in a number of projects while working as partner or consultant to ship owners and designers. Figure 2 shows a number of such "real world" projects in which the method has its roots.
- Predictions have been correlated to full-scale results [7].
- The method uses the same modelling approach to evaluate different WPS. This provides customers with decision support material that evaluates competing WPS in a fair wayOnce the business case for a particular WPS has been made and as the ship design process progresses, the input to the analysis model can gradually be refined. Database values can e.g., be replaced with ship specific CFD or model test results.

3.2 SHIP PERFORMANCE MODULE

At the core of the 'ship performance' module from Figure 3 is a power prediction based on the modified ITTC-1978 approach described in [8]. The prediction is carried out in SSPA's inhouse software suit SEAMAN and balances hull and sail forces in 4 DOF (surge, sway, roll, yaw) [9, 10].
Input to the simulations is a 180 x 25 matrix of true wind angle (TWA) and true wind speeds (TWS) combinations. For each of these SEAMAN carries out a performance prediction and returns a vector with parameters like: Sail forces, ship speed, required shaft power, propeller *rpm*, leeway angle, rudder angle, heel etc. This output is stored in the format of 'polars' for further analysis in the 'voyage simulation' module.
The individual force components that SEAMAN balances when computing the polars are modelled as follows:

3.2.(a) Resistance modelling

Calm water resistance can be modelled based on generic methods (e.g. Holtrop Mennen [11]) but is preferably taken from our database of model test results for similar ships.
Wind resistance of the ship's above water part is taken from SSPA's database of wind tunnel test results.
Added resistance in waves, for vessels at an early concept stage, is calculated based on a modified "STAWAVE-2" model [12] and spectral superposition theory. A formulation based on the more universal model proposed by Liu & Papanikolaou [13] is currently being implemented. Alternatively, added resistance can be taken from SSPA's database.
A "waves follow wind" approach is used to work out wave heights and wave spectra. The exact wind-wave relationship depends on the operational area. For, e.g., the North Atlantic relationships according to [14] are used.

Figure 2: Selected recent wind ship projects in which SSPA has been involved.

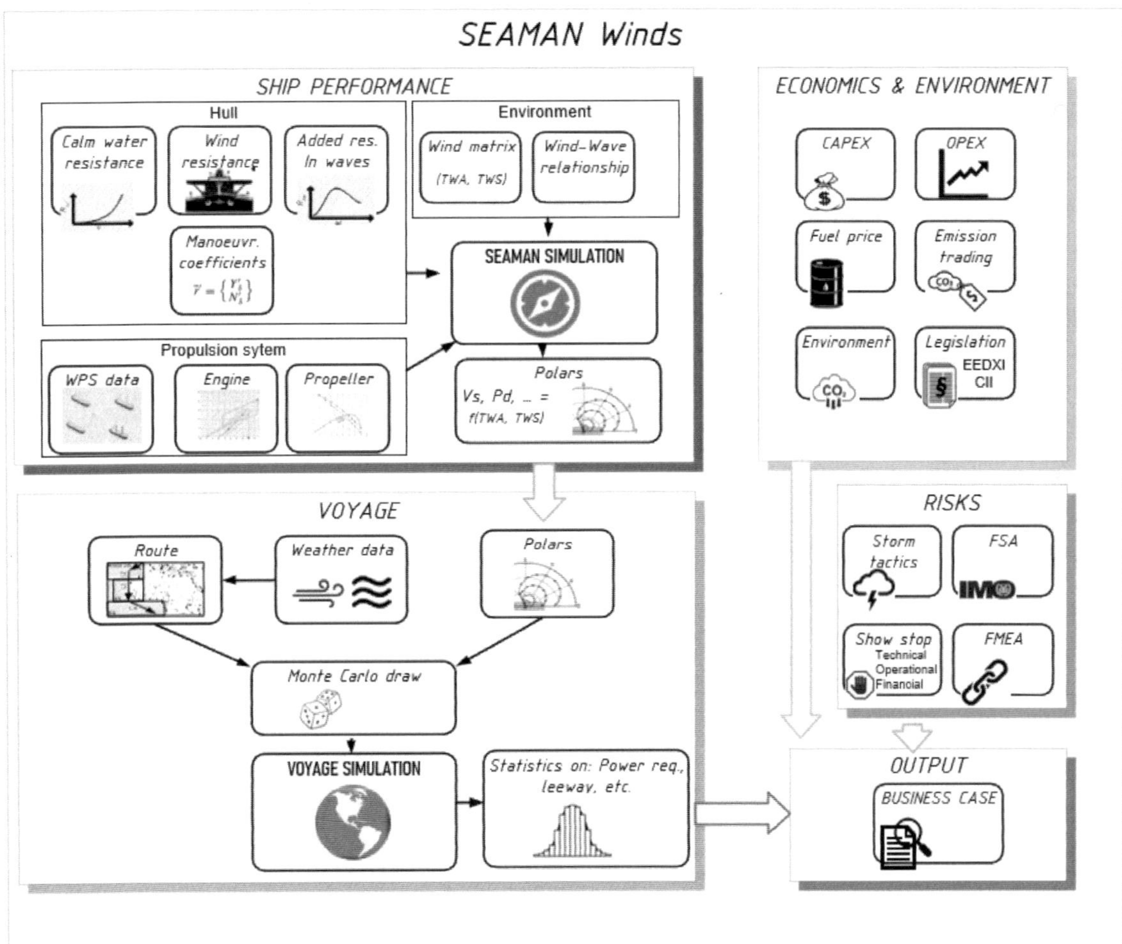

Figure 3: Block diagram illustrating the methodology behind *SEAMAN Winds*

3.2.(b) Manoeuvring forces

Hydrodynamic derivatives are extracted from SSPAs database of manoeuvring model tests. Regression is used to find a best fit of derivatives based on the main particulars of the vessels. At a later stage in the design spiral, these can be replaced by ship specific values from CFD calculations and/or model test.
Force modelling includes induced resistance due to leeway and rudder deflection as well as propeller efficiency due to angled inflow.

3.2.(c) Conventional propulsion system

Propeller characteristics are modelled based on open water curves and propulsive factors from SSPA's database.

Engine characteristics can be modelled based on the generic power-*rpm* relationships provided in [15]. The only required input is Maximum Continuous Rating (MCR) and the corresponding *rpm*.

3.2.(d) Wind propulsion system

An important part of *SEAMAN Winds* is the module that calculates the aerodynamic forces from the WPS.

Rotor sails are modelled as described in [7]. Generic lift and drag curves for a number of rotors with the usual aspect ratios found on today's market have been derived using CFD, see [16] for more information and validation cases. In the CFD simulations the rotor is placed on a symmetry/mirror plane.
For ships with several rotors *SEAMAN Winds* then models 3D rotor-rotor interaction based on a generalised model derived by CFD. Hull-rotor interaction is modelled in a similar way, see [7] for more details.
The operational *rpm* of rotors depends on wind speed and direction. This relationship can be modelled based on tabulated values from rotor suppliers or generic calculations. The electrical power required to drive the rotors is derived in a similar way and subtracted off the power savings provided by the WPS.

Wingsails are modelled using the CORR-SILL-method described in [17, 18]. This method is based on combining 2D sectional lift and drag data from RANSE calculations with lifting line theory to account for 3D effects like induced drag and mutual interaction between multiple wingsails, Figure 4.

The Kite module is based on the work carried out at SSPA during the EffShip project in 2012 [19]. It leans on the ideas presented in [20, 21] and assumes that the kites flies in a figure-eight pattern on a quarter-sphere whose radius is equal to the length of the towing line.

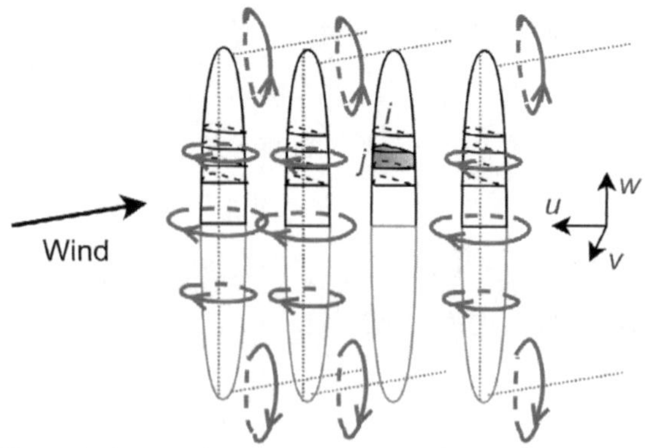

Figure 4: Assumed vortex system used in CORR_SILL to account for interaction effects between multiple wingsails, from [18].

3.3 VOYAGE SIMULATION MODULE

As illustrated on the lower left-hand side of Figure 3 the output of the SEAMAN simulations from the ship performance module is analysed in the voyage module where information about the intended trading route is combined with weather statistics and the SEAMAN-polars.
More specifically the route is divided into a number of legs. For each of these, a discrete joint weather distribution for true wind speeds (TWS) and true wind angles (TWA) is aggregated based on wind data from the ERA5 reanalysis dataset available at the Copernicus Climate Data Store [22]. Figure 5 shows an example.
Each leg is treated independently, and leg-wise distributions are assumed to be uncorrelated.
To keep the computational effort within reasonable limits, a Monte Carlo simulation is then performed according to the following steps:

1. For each of the typically $100 \cdot 10^3$ iterations in the Monte Carlo simulation
   a. For each leg on the route
      i. Find the course over ground and distance that the ship will travel on this leg
      ii. Draw a sample weather condition from the discrete, joint weather distribution. The sample is randomly chosen based on its probability of occurrence, i.e. a weather condition with 2% probability of occurring has a 2% probability of being sampled.
      iii. Evaluate ship performance for this specific weather condition using the performance polar curves from SEAMAN, linearly interpolating where necessary.
   b. Aggregate the results for the entire route based on the sub-results from each leg.
2. Combine the aggregate route results from all iterations into probability distributions

The output from the voyage simulations is therefore probability density functions for all the parameters that were 'sent into' the Monte Carlo simulations, e.g. sail forces, ship speed, required shaft power, propeller *rpm*, leeway angle, rudder angle, heel etc. The distributions can be analysed to answer questions like:

- What will the average power savings on this route be?
- What is the 95 % confidence interval of achieving a power saving of 5 %?
- What is the probability of exceeding sail forces of X Newtons or rudder angles of Y° while trying to stay on course?

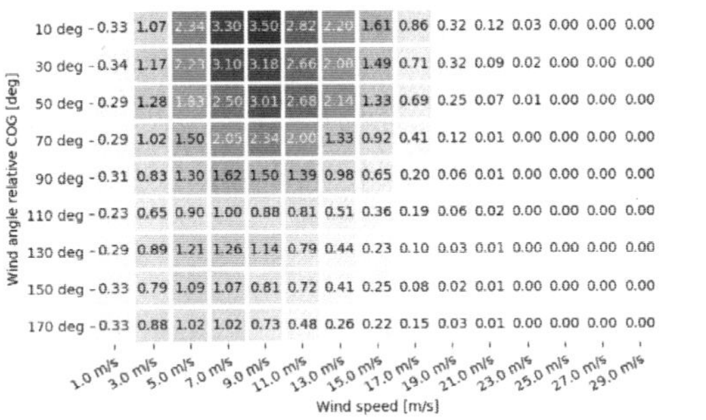

Figure 5: Discrete, joint probability distribution of TWS and TWA, 10 m above the sea surface. Numbers in boxes are probability expressed in %, TWA relative Course over Ground (COG), 0° is headwind, from [7].

3.4 ECONOMICS & ENVIRONMENTAL MODULE

The 'economics and environmental module' of *SEAMAN Winds* estimates investment costs (CAPEX) and maintenance costs (OPEX) for wind propulsion systems and compares these to power (and fuel) savings from the voyage simulations.
The module also uses the information from the voyage simulations to analyse the effect of the WPS on CO_2 emissions and regulatory parameters like EEDI, EEXI or CII.
Based on such calculations, financial information on e.g., the payback time for the installation of a WPS or the effect of measures like 'carbon trading' can be derived.

Inside the module WPS costs can be modelled in different ways. For the more established, commercially available systems, supplier information and open-source data can be used e.g. [23, 24.]. Table 1 summarises e.g., how CAPEX and OPEX of kite sails are modelled.

Table 1: Modelling the costs of kites, in 2016 USD, [24].

Size of kite [m²]	Purchase costs [k$]	Installation [% pur. costs]	OPEX [% pur. costs]
160	280	7.5%	4%
320	480		6%
640	920		8%
1280	1755		10%
2500	2590		12%
5000	3430		14%

For innovative systems that are not yet available of-the-shelf costs are more difficult to estimate and a bottom-up life-cycle-cost approach has to be taken. The method used here is based on breaking down the proposed WPS into its components, use "triple-estimates" to estimate component costs, refine the breakdown structure where required, and sum the results to work out total costs [25].

3.5 RISK MODULE

A unique feature of *SEAMAN Winds* is the option to assess risks associated with installing a WPS. Such risks fall into the categories 'technical', 'operational', and 'financial'. They range from mechanical failure of the WPS over crew training and recruitment issues to the consequences of a year with very little wind.
One of the methods used in the module is that of the Formal Safety Assessment (FSA) as outlined by IMO. It consists of five basic steps [26]:

1. Hazard identification - list of accident scenarios
2. Risk analysis - probability and consequences
3. Risk control options
4. Cost benefit assessment
5. Recommendations

During step 1 all hazards related to technical, regulative, external factors (extreme weather), and human factors are identified based on a rough tentative WPS design. Subsequently the hazards are ranked in terms of their probability and severity of consequences. Such an analysis provides guidance on prioritization on issues to be subject for more detailed risk analyses. The method identifies and addresses all technical and operational hazards, their causes, potential consequences, and potential preventive and mitigating safety barriers by applying a structured approach, including various root causes and safety barriers along the causal chain, see Figure 6.

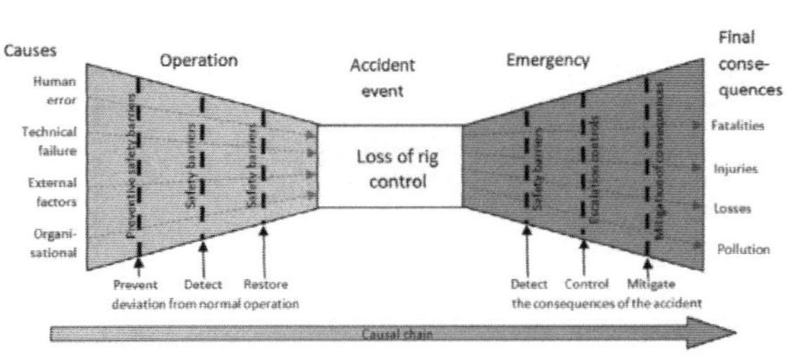

Figure 6: Bow-tie risk analysis approach.

Further on in the design process of the ship, and particularly in the case of innovative WPS, classification societies might require more detailed investigations like 'Failure Mode and Effect Analyses' (FMEA). Such an analysis can address i.a. the possibility to reduce sail area in rough weather. An FMEA identifies if a single failure in any of the WPU components may impede the reefing capability, thus resulting in recommendations which rig components and systems need to be redundant.

4. EXAMPLE APPLICATION

In this section the methods described above are used to study the case of the RoPax ferry m/v *Copenhagen*, operated by Scandlines between Gedser in Denmark and Rostock in Germany. The vessel (L=156.45m, B=24.6m, IMO 9587867) has been equipped with a 5m x 30m Norsepower rotor sail with an end plate. The WPS is positioned longitudinally around mid-ship, 17.2 m above design water line, see Figure 7 and [7] for more details.

Figure 7: M/V *Copenhagen* with 5x30m rotor sail, illustration Scandlines.

4.1. RESULT EXAMPLES ROTOR SAILS

A few typical results from the individual modules of *SEAMAN Winds* are shown and discussed below.

4.1 (a) Results from the ship performance module

Figure 8 shows an example polar plot that was calculated by the ship performance module. The diagram shows the power reduction due to the rotor relative to the baseline case without the WPS. The numerical model behind this plot has been correlated against full-scale sea trials, [7].
As expected, the most favourable points of sailing with a rotor are beam reaching and broad reaching, compare e.g. [27].
For very low true wind angles TWA<≈ 20° power savings are negative. This reflects the drag of the ideling rotor in headwind.
Under optimal conditions (TWS≈12m/s TWA≈90° power savings can reach up to 28%.

4.1 (b) Results from the voyage module

As the vessel will experience different wind and wave conditions over time, the power for a constant service speed will differ from journey to journey.
These effects are evaluated in the voyage module of *SEAMAN Winds*, which returns statistical information on power savings, CO_2 reductions etc. Figure 9 shows an example for *Copenhagen* sailing at a constant speed of 16 knots.
The figure displays the probability distribution of the required shaft power for cases with and without rotor.
Power requirements 'under sail' depend on the wind and therefore the 'with rotor' curve exhibits the larger spread. Differences between the curves are largest at the lower power range (to the left in the figure).

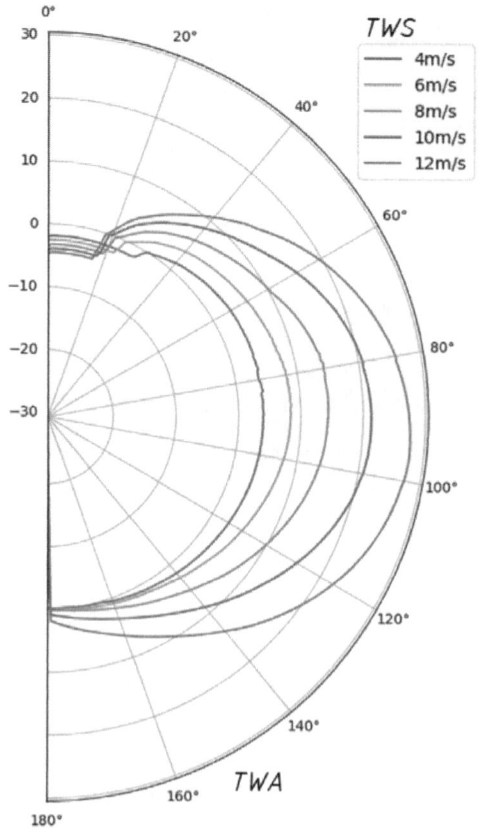

Figure 8: Net power savings due to installation of Flettner rotor Fixed ship speed of 16 knots.

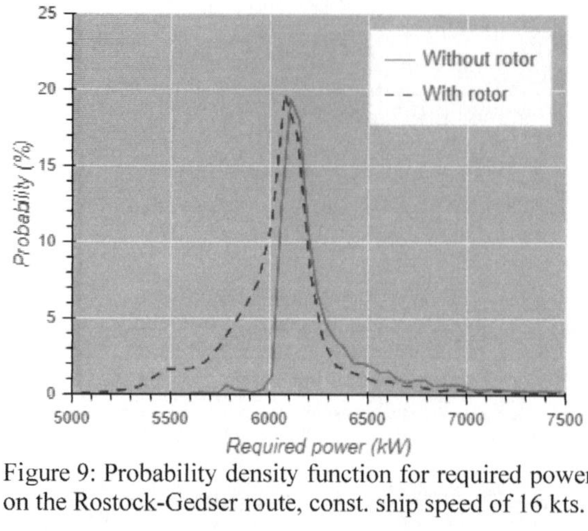

Figure 9: Probability density function for required power on the Rostock-Gedser route, const. ship speed of 16 kts.

This will be due to the fact, that the most common wind direction on the North-South route is westerly. Such beam winds are ideal for the rotor, but also not very harmful for the case without rotor. The higher power range (to the right in the figure) covers the days of strong head wind, when both cases struggle, compare also [7].

Averaged over a long period, the difference in power requirement corresponds to a 3.9% reduction in power. This already takes the power required to operate the rotor into account.

Similar probability density functions can be extracted for all the other variables analysed in the ship performance module (rotor forces, heel, rudder, and leeway angles etc.)

4.1 (c) Results from the economics and environmental module

Figure 10 and Table 2 illustrate typical output from the economics and environmental module. The numbers and plots only serve to illustrate how *SEAMAN Winds* works. They are estimates and do not represent actual costs or how Scandlines operates the vessel.

Table 2: Example analysis.

Input parameter	Value
Average power saving on route	250 kW
Number of return journeys per day	5
Operational days per year	350
Main engine specific fuel oil consumption	200g/kWh
Output parameter	**Value**
Energy saved per year	1 240 MWh
Fuel savings per year	248 t
Bunker savings (@510$/t) VLSFO	100 000 €/a
CO_2 savings per year	771 t

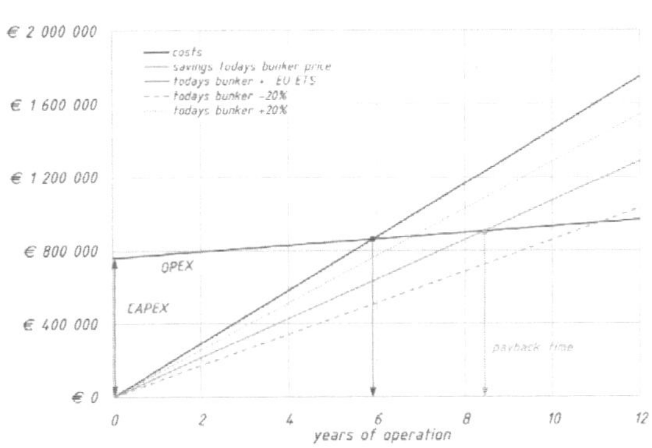

Figure 10: Costs and saving development over time. Numbers are for illustration only and do not reflect how vessel is operated in reality.

As can be seen in Figure 10, payback time for the WPS in this generic example ranges from 7 to 11 years, depending on fuel prices.

Including maritime transport in the European Union Emissions Trading System (EU ETS) would, at todays 'carbon price' of 50€/t, earn an additional 38 000 € in revenue per year. As illustrated in the figure this has a huge influence on payback time, decreasing it from 8.5 to 6 years.

The predicted average CO_2 savings from Table 2 are also an important finding and can be used to assess the impact of the Flettner installation on compliance with EEDI/EEXI rules or the CII rating scheme that will enter into force in 2023.

5. CONCLUSIONS: HORSES FOR COURSES

Wind propulsion systems (WPS) are a promising way to reduce CO_2 emissions from shipping. Among the key barriers to the market-uptake of such systems are high investment costs combined with uncertainty around how effective WPS are in reality and what risks they entail.

In this paper we present a systematic, scientific methodology to address these issues and help shipowners and operators make informed decisions when looking at the business case of installing a WPS. The proposed method was developed into a Software called *SEAMAN Winds*. This tool can be used to systematically assess the benefits and drawbacks of different WPS technologies.

The vast majority of the prediction and assessment methods used in *SEAMAN Winds* is not new but has been used before, both, by SSPA and other institutions. What is new, is the integration of these time-proven methods into one single, streamlined process. Unique features include:

- Hull and propeller modelling is based on SSPA's large database of experimental and CFD results.
- It has been developed in dialog with end-users from the industry and has a commercial rather than an academic or research focus.
- It can assess the risks associated with WPS installations.
- The method uses the same modelling approach to evaluate different WPS. This provides customers with decision support material that evaluates competing WPS in a fair way. Once the business case for a particular WPS has been made and as the ship design process progresses, the input to the analysis model can gradually be refined with increasingly complex and accurate simulation input.

When it comes to making the business case the method can answer questions like:

- Which WPS technology is best for 'my' route?
- What if we change route?
- What if we change speed?
- What if we increase WPS size?
- What if fuel prices increase 10% each year?
- How much will it help with EEXI and CII?
- How will policy changes (e.g. emissions trading) affect the business case?

Using the example of a RoPax ferry with a rotor sail it is shown, how the methods works and what kind of decision support material it can produce. It is calculated, how many tonnes of CO_2 can be avoided each year and how payback time is affected by different factors.

Table 3: Some of the aspects to consider when selecting a WPS

	Aspect	Flettner rotors	Kite	Wingsails	Soft sails	Suction wing
Maturity	**Proven technology**	*First tested in the 1920s. Most accepted + proven solution on the modern market.*	*Tested in a few commercial vessels*	*Some experience from applications in the 1980s* *Yacht racing*	*Long history of traditional sails.*	*Some experience of Turbosails. New technology being tested.*
Commercial	**System/ equipment availability**	*Commercially available*	*Commercially available*	*Several concepts Only a few of-the-shelf.*	*Available in large scale super yacht segment.*	*Available in small scale.*
	CAPEX	*High*	*Low*	*High*	*Medium*	*High*
	OPEX	*Low*	*High (wear and tear)*	*Low, medium if reefable*	*High (wear and tear)*	*Low*
Technical	**Retrofit off hire impact**	*Weeks*	*Days-Weeks*	*Weeks-Month*	*Weeks-Month*	*Hours (containerised small-scale units)*
	Deck space requirements	*Open deck for optimal performance.*	*Deployed form a small mast near the bow.*	*Open deck for optimal performance.*	*Open deck for optimal performance.*	*Removable container can be placed on cargo holds.*
	Operational wind directions	*Optimal operation in 90° - 150°).*	*Broad reach to downwind. Risk that kite falls into the water in low wind speeds.*	*Efficient upwind, less efficient downwind.*	*Similar to wingsail but less efficient.*	*Similar to wingsail but more efficient.*
	Sail trimming	*Only rpm can be adjusted.*	*Figure-eight pattern can be adjusted to windspeed.*	*Not many options to trim rigid wings. Twist impossible.*	*Can be trimmed in many ways.*	*Not many options to trim rigid wings. Twist impossible.*
Operational	**Cargo handling constraints**	*Foldable and movable solutions available on the market.*	*None*	*Some concepts are foldable or collapsible to avoid constraints.*	*The rig implies constrains even with furlable and reefable sails.*	*Removable container available.*
	Air draft	*Largest available systems today 35m and tiltable*	*No constrains when not sailing*	*Often taller than rotor sails.*	*Often taller than rotor sails.*	*Containerised system 10.5 m.*
	Risks imposed by adverse weather	*Slender profile and robust design*	*Can be retrieved in storm*	*Require fast vaning response*	*Reefable sails*	*Small systems collapsible*
	Operational effects	*Risk of reduced visibility. Gyroscopic roll stabilization effect. Reverse rotation possible for improved* manoeuvrability.	*No contribution to heel. Cannot be used in areas with dense traffic.*	*Risk of reduced visibility.*	*Risk of reduced visibility.*	*Risk of reduced visibility.*

6. REFERENCES

[1] T. CHOU, V. KOSMAS, K. RENKEN, M. ACCIARO, 'New Wind Propulsion Technology - A Literature Review of Recent Adoptions,' Interreg North Sea Europe Wind Assisted Ship Propulsion (WASP) report D 5.B, 2020.

[2] BOUMAN, E., LINDSTAD, E., RIALLAND, A., STRØMMAN, A., 'State-of-the-art technologies, measures, and potential for reducing GHG emissions from shipping – A review', *Transportation Research* Part D: Transport and Environment, Volume 52, Part A, 2017.

[3] INTERNATIONAL WINDSHIP ASSOC., Technology and Design, *https://wind-ship.org/en/technology-design*, retrieved August 2021.

[4] TILLIG; F., 'Simulation model of a ship's energy performance and transportation costs', *PhD thesis*, Chalmers University of Technology, 2020.

[5] MARIN, 'Blue Route' https://blueroute.application.marin.nl, retrieved August 2021.

[6] BAR TECHNOLOGIES, 'Ship System Efficiency Analysis Toolset', *www.bartechnologies.uk /project/shipseat*, retrieved August 2021.

[7] WERNER, S., NISBET, J., HÖRTEBORN, A., NIELSEN, R., 'Speed Trial Verification for a Wind Assisted Ship', *Wind Propulsion 2021*, RINA, 2021.

[8] GERHARDT, F. C., KJELLBERG, M., 'Determining the EEDI weather factor fw', *Influence of EEDI on Ship Design & Operation*, RINA, 2017.

[9] SSPA SWEDEN AB, www.sspa.se/tools-and-methods/seaman-simulations, retrieved August 2021.

[10] OLSSON, F., MARIMON GIOVANNETTI, L., WERNER, S., FINNSGÅRD, C., A Performance Depowering Investigation for Wind Powered Cargo Ships Along a Route', *InnovSail 2020.*

[11] HOLTROP, J., MENNEN, G. G. J., 'An approximate power prediction method', *International Shipbuilding Progress*. 29, 1982.

[12] ITTC Recommended Procedure and Guideline 7.5-04-01-01.1, 'Preparation, Conduct and Analysis of Speed/Power Trials', 2021.

[13] LIU, S., PAPANIKOLAOU, A., 'Regression analysis of experimental data for added resistance in waves of arbitrary heading and development of a semi-empirical formula', *Ocean Engineering*, Vol. 206, 2020,

[14] BALES; S .L., 'Designing Ships to the Natural Environment', *Naval Engineering J.*, vol. 92(2) 1983.

[15] MAN, 'Basic Principles of Ship Propulsion', brochure available online, retrieved 2019.

[16] LI, D.-Q., LEER-ANDERSEN, M., ALLENSTRÖM, B., 'Performance and vortex formation of Flettner Rotors at high Reynolds numbers', *29th Symposium on Naval Hydrodynamics,* Gothenburg, 2012.

[17] 8. PERSSON, A., Li, D.-Q., OLSSON, F., WERNER, S., 'Performance predictions of wind propulsion systems using 3D CFD and route simulations,' *International Conference on Wind Propulsion*, RINA, 2019.

[18] MALMEK, K., DHOMÉ, U., LARSSON, L., WERNER, S., RINGSBERG, J., FINNSGÅRD, C., 'Comparison of two rapid numerical methods for predicting the performance of multiple rigid wing-sails', *5th International Conference on Innovation in High Performance Sailing Yachts and Sail-Assisted Ship Propulsion*, 2020.

[19] EFFSHIP project, *www.sspa.se/environment-and-energy efficiency/effship-project*, retrieved August 2021.

[20] P. NAAIJEN, P., KOSTER, V., DALLINGA, R. P., 'On the power savings by an auxiliary kite propulsion system', *International Shipbuilding Progress* 53, 2006.

[21] TRAUT, M., GILBERT, P., WALSH, C., BOWS, A., FILIPPONE, A., STANSBY, P., WOOD, R., 'Propulsive power contribution of a kite and a Flettner rotor on selected shipping routes', *Applied Energy*, vol. 113, 2014.

[22] COPERNICUS DATA STORE, *https://cds.climate.copernicus.eu/#!/home* , retrieved August 2021.

[23] NY TIMES MAGAZINE, *//www.nytimes.com/2021/06/24/magazine/ cargo-ships-emissions.html*, retrieved August 2021

[24] DNV_GL, 'EE Appraisal Tool for IMO, project report, DNV-GL report 2015-0823,Rev.0, 2016.

[25] LICHTENBERG, S., 'Proactive management of uncertainty using the successive principle: a

practical way to manage opportunities and risks', *Polyteknisk Press*, 2000.

[26] IMO, 'Revised Guidelines for Formal Safety Assessment (FSA) for Use in the IMO Rule-Making Process', IMO MSC-MEPC.2/Circ.12/Rev.2., 2018

[27] RECHE-VILANOVA, M., HANSEN, H., BINGHAM, H., 'Performance Prediction Program for Wind-Assisted Cargo Ships', *Journal of Sailing Technology*, vol. 6(1), SMAME, 2021.

7. AUTHORS BIOGRAPHY

Frederik C. Gerhardt is Senior Technical Specialist for seakeeping & ship performance in waves and aerodynamics at SSPA. He is responsible for developing SSPA's product portfolio within the area of wind propulsion. He chairs the 30th ITTC seakeeping committee.

Sofia Werner is responsible for method development within fluid dynamics at SSPA. Her previous experience includes towing tank testing and ship design using CFD. 2011-2017 she was a member of the ITTC committee that developed the Recommended Procedures for Speed Trial. She chairs the ITTC Specialist Committee for Wind Powered Ships.

Axel Hörteborn holds the current position of Specialist at SSPA. He is responsible for ship simulations, AIS data and Monte Carlo simulations analysis. His previous experience includes risk assessments of large-scale Monte Carlo inspired ship simulations and simulations of sailing vessels.

Olov Lundbäck is Senior Specialist within Simulations and Algorithms at SSPA. He has a Ph.D. in Seakeeping and Manoeuvring of High Speed Craft and has technical responsibility for Ship Motion Simulations.

Jonny Nisbet holds a position as Senior Specialist at SSPA. His previous experience includes manoeuvring and seakeeping model testing, as well as simulation and CFD.

Tobias Olsson is Project Manager at SSPA with a background in marine engineering and maritime management. His main focus is life cycle modelling.

ESTIMATING THE COST OF GREENING OF SHORTSEA VESSELS AND THE MARKET POTENTIAL OF GREEN RETROFIT

O Schinas, HHX.blue, Germany

SUMMARY

Shortsea shipping (SSS) has attracted the interest of re-searchers and policy-makers as a viable transport alternative against mainly road haulage. Despite policy initiatives, SSS has not increased its market share nor attracted fresh interest in the market, so the fleet is ageing. Lately, due to the de-carbonization effort of the maritime industry and generally of greening the economy, SSS becomes an interesting trans-port alternative that deserves further research. This work fo-cuses on the cost of decarbonising the global SSS fleet with a parallel interest on the EU fleet. This cost estimation is based on recent publications, and considers ships expected to be ordered and demolished; for the existing fleet the cost of retrofitting with greening technology is also estimated. Based on reasonable assumptions, the potential market of wind-assisted technologies is estimated too. This work is an outcome of the WASP Interreg project.

1. INTRODUCTION

Shortsea shipping (SSS) can be defined as any transport by sea that does not involve ocean crossing. In this regard, usu-ally ships of less than 30k deadweight tons (DWT) are en-gaged in these trades; most passenger roro ferries are prac-tically classified as SSS vessels, while many dry and wet cargo ships offer their services in local, regional and coastal trades. The subject of SSS is extensively discussed in the literature [see, e.g., 13, 20] The interest on SSS in the European Union (EU) dates back to 1992, when the Common Transport Policy promoted poli-cies to shift cargoes from the road to the sea and rail, due to congestion, financial needs and environmental performance [4]. The importance of SSS is reflected in the statistics pro-vided by EUROSTAT for 2018, see Figures 2, 2, and 2. Ap-parently, SSS enjoys a high market share in terms of tonnes, as almost 60% of the freight is moved by SSS means (Fig-ure 2); hence the strategic importance of efficient and green SSS transportation assets is irrefutable. Furthermore, most of the assets are engaged in liquid bulk, containerised and roro trades (Figure 2); this is a critical aspect when con-sidering fleet renewal either from a financial or policy per-spective. Finally, the regional importance of SSS is demon-strated in Figure 2, where the Black Sea, North Sea and Baltic trades enjoy the highest market shares; moreover, this figure shows also the critical importance of SSS in achieving or retaining social cohesion among regions of the EU. The SSS fleet is generally overaged, and many SSS ves-sels exceed reasonable limits of technical and economic life.

DWT range	wet	dry	con-ships	general cargo
0-5k	13.6		20.9	27.6
5-10k	15.6		18.3	
10-15k	27.6	14.6	13.3	

Table 1: Average age of SSS assets, Source: Own analysis based on published data [21]

However, this reflects the lack of competitive advantages of newer tonnage. Younger ships are usually mortgaged and capital expenses (CAPEX) are deteriorating the available margins. As the market is well developed, the service is not differentiated, and perfect competition prevails, the profit margins are generally thin and cannot support further fleet renewal plans, unless older assets finance with their resid-ual value the acquisition of younger ships or the ordering of newbuildings. In most cases, these ships remain under the radar of most financial houses, due their usual spot employ-ment, market values, and ticket sizes. In addition to the market inherent difficulties in securing capital, despite its critical importance in global and Euro-pean trades, owners have to face the requirement of green-ing the assets. Nevertheless, this greening requirement may accelerate both the renewal of the SSS fleet as well as the retrofit of existing ships by installing greening technologies, such as wind-assisted propulsion (WASP). From a regula-tory perspective, ships of various sizes and type are regu-larly dry-docked as a result of the statutory requirement of SOLAS Ch-I Reg.10, therefore the installation of greening technologies, such as WASP may be effectively scheduled when the ship has to undergo the statutory survey. The decarbonization of marine operations and generally the abatement of Green-House Gases (GHG) is currently in progress, as the International Maritime Organization (IMO) adopted the Energy Efficiency Index (EEDI) and consid-ers further measures to reduce CO_2 emissions [9, 7, 8]. The policy framework on air quality as well as its im-plications are widely discussed in the literature [see e.g. 12, 16]. Recent discussions on potential technical, oper-ational (e.g. speed limits) and financial measures, the so called Market-Based Measures (MBM), the Marine Envi-ronment Protection Committee (MEPC) in its 75th session in November 2020 approved amendments to MARPOL An-nex VI proposed in October by the IMOs Intersessional Working Group on Reduction of GHG. These amendments are briefly outlined below:

1. The Energy Efficiency Existing Ship Index (EEXI) is introduced from 2023. Practically every ship should

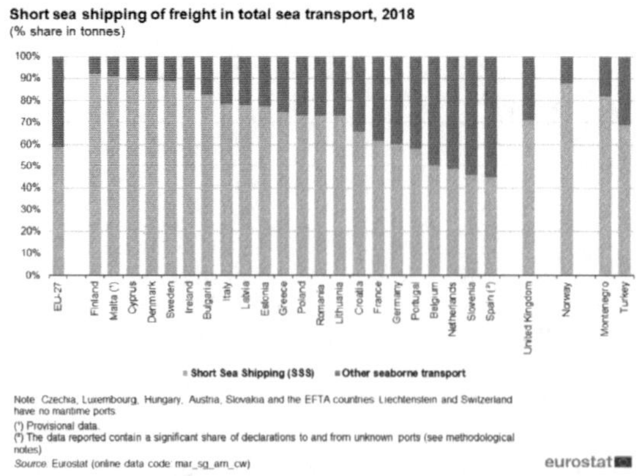

Figure 1: A closer look at the importance of SSS in the EU, Source: [5]

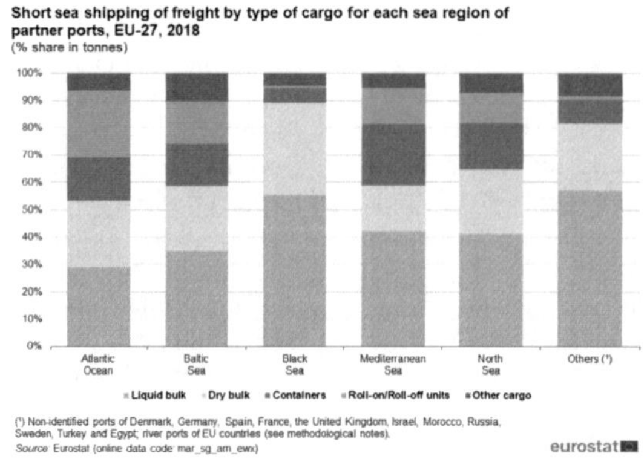

Figure 2: SSS Cargo Breakdown, Source: [5]

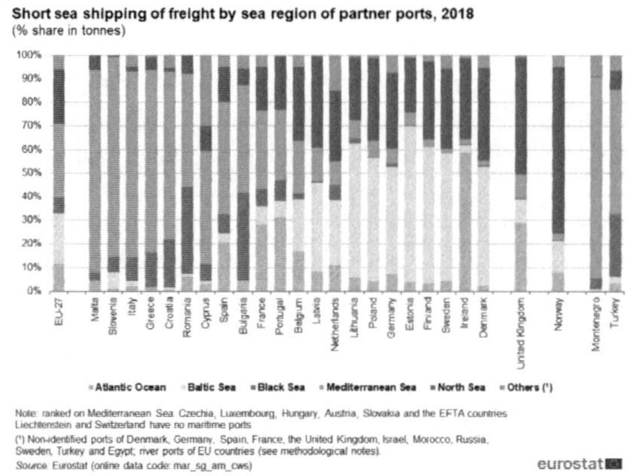

Figure 3: SSS Cargo Breakdown, Source: [5]

reach a level below the baseline, which is currently equivalent to the reference lines of EEDI Phase 2 or Phase 3 (from 2023/25). Owners can meet this require-ment through a range of methods, such as engine power limitation, using alternative fuels, or retrofitting with energy saving technologies (EST), such as the wind as-sisted propulsion (WASP) technologies. Recent pub-lications provide a thorough overview of WASP tech-nologies and their current status [2, 15].

2. The Carbon Intensity Indicator (expected to be the An-nual Efficiency Ratio (AER) for most ships) is intro-duced from 2023. All vessels will be required to calcu-late their AER and they will be allocated a rating on an annual basis. This rating will vary from A to E, with C indicating that a vessel is reducing its carbon inten-sity in line with the IMOs intended pathway to reach the global 2030 target; vessels with a rating of D (for 3 years) or E (at any point) will be required to submit improvement plans.

3. The EEDI Phase 3 will be brought forward from 2025 to April 2022 for vessel types including containerships, most gas carriers, general cargo ships and cruise ships. Reductions of 30 – 50% from the EEDI baseline will be required under Phase 3 for containerships.

The above measures highlight the regulatory uncertainty that increases operational and financial uncertainty in tandem. Measures like EEXI and AER lead to the question if a given ship is capable to trade as is or a green retrofit is necessary; the latter question implies also if retrofit is financial mean-ingful or nor. The decision to bring forward EEDI Phases might also have a positive impact as the interplay of asset prices and environmental regulation and the bias (or strate-gic option) of lenders to finance greener tonnage is proven [14]. The cost of retrofitting as well as the analysis of the market potential has not received special attention from researchers, due to the inherent problem of data collection that jeopar-dises any effort for a systematic approach. Furthermore, the need of greening marine operations dictates not only the renewal of the fleet, in terms of demolition and in-creased order-book of newbuildings, but also the installa-tion of greening technologies - retrofit- on existing vessels. Although these topics are not sufficiently covered in the lit-erature, a recent publication estimates the financial require-ments for the greening of the global fleet [17]; this work pro-vides a solid basis for further analysis in this paper. In addi-tion, reported retrofit data are presented in Table 2; all these cases refer to installation of greening technologies, such as WASP, or even larger conversions, such as of new bows or construction of new fuel tanks and outfitting for dual fuel systems. This table is a valuable tool The rest of the paper is structured as following: in the next Section 3 the elaboration of recent data and results estimat-ing the financial needs of the SSS fleet will be discussed. The analysis will focus on retrofits, as the new building markets are impacted by long-term policies and intertwined global economic strategies. Then in Section 4, the potential of the market per fleet segment for WASP technologies will be approximated based on reasonable yet averaged assump-tions. The last Section 5 summarises the key findings. Last but not least, the main motivation of conducting the above research, is the need to address better the requests of SSS owners on using WASP technologies; this work is a product of the effort within the North-Sea Region Interreg WASP Project.

2. FINANCIAL NEEDS FOR GREENING

The reported cost (nominal investment in USD) of Table 2 reflects the wide spectrum of technical options available, but also the lack of transparency and sufficient statistics to ex-tract the unit cost of USD/DWT; ideally one could relate the extracted unit cost with achieved -reported- reduction, so the marginal abatement cost could be approximated, subject to reasonable assumptions. Based on recent results of a bottom-to-top analysis [17], along with further elaboration of the published data, Tables 3 and 3 as well as Figure 3 are derived. The original results of Schinas and Bergmann [17] do not consider a clustering of SSS vessels, therefore the rather hard threshold of 30k DWT was considered; all data used or presented below refer to the $< 30k$DWT cluster of the dataset and results of this work. Moreover, this work focuses on short-term financial needs, in particular till 2025-6, due to the limitations in estimating the order book of newbuildings in the global market.
The analysis of Table 3 yields the following interesting re-sults:

1. Roro and passenger ships seem to have a higher finan-cial need;

2. tankers, gas carriers and containerships seem also also in need for finance;

3. the short-term need for newbuildings is circa USD30bn, i.e. almost 10% of the current top-40 banks exposure to shipping [19]; roro-passenger ships amount circa 50% of this sum;

4. the short-term market for retrofit amounts circa USD11bn;

5. bulkers seem to have the greater potential for retrofit.

The last column of Table 3 illustrates the financial need as estimated over the total financial need of the global fleet, as approximated by Schinas and Bergmann [17]; in this publication the total financial result considers the revenues from demolishing older tonnage, an aspect that is not taken into account in the SSS estimation (newbuilding+retrofit-demolition=total financial need [Eq.2,p.2 17]). Therefore, some of the final results are very high, such as in the case of reefers, general cargo and roro-passanger ships, where these segments are in statistical decline in terms of number of ships in service. The above results should be critically assessed as the sums are approximated by deducting the < 30kDWT ships

Source	Measure	Vessel	Nominal investment USD	Implied USD/DWT	Reduction Potential
[1]	Dual fuel engine switch	n.a.	15,000,000	n.a.	2.0 - 4.0%
[2]	Flettner rotor	Various implementations	n.a.	n.a.	0.4 - 50.0%
	Kites	Various implementations	n.a.	n.a.	1.0 - 50.0%
	Wing sails	Various implementations	n.a.	n.a.	5.0 - 60.0%
[3]	Bulbous bow optimization	4860 TEU Container ship (approx. 65k DWT)	600,000	9.23	6.0%
[6]	Waste heat recovery	120-200k DWT Tanker	3,512,550	21.95	6.0 - 8.0%
	Towing Kites	10-20k DWT Chemical Tanker	987,100	65.80	4.6 - 9.2%
	Solar energy	5-10k DWT Products Tanker	1,330,000	177.33	1.2%
	Integrated propeller and rudder	5-10k DWT General Cargo	415,360	55.38	2.0 - 6.0%
[10]	Retrofit power plant	2,350 DWT Ro-Ro cargo ship	n.a.	n.a.	n.a.
[11]	Flettner rotor	9,700 DWT Ro-Ro vessel	n.a.	n.a.	5.0 %
[23]	Flettner rotor	109,647 DWT Product Tanker	n.a.	n.a.	8.2%
[22]	Bundle of fuel efficiency retrofit technologies	46k DWT Bulk Carrier	1,000,000	21.73	n.a.
	Dual fuel engine switch	46k DWT Bulk Carrier	10,000,000	217.39	n.a.

Table 2: Reported Retrofit Data

	Road	Rail	Inland Waterways	Pipelines	Shortsea	Air	Total[1]
1995	47%	16%	5%	4%	28%	0%	*2400*
2000	49%	14%	5%	4%	28%	0%	*2753*
2005	51%	13%	4%	4%	27%	0%	*3088*
2010	52%	12%	5%	4%	27%	0%	*3026*
2015	51%	13%	5%	3%	28%	0%	*3072*
2016	51%	13%	5%	3%	28%	0%	*3193*
2017	52%	13%	4%	3%	28%	0%	*3294*
2018	51%	13%	4%	3%	29%	0%	*3353*
y-o-y							
"95-18"	1.75%	0.51%	0.43%	0.01%	1.58%	1.88%	1.40%
"08-18"	0.17%	0.04%	-0.77%	-0.89%	1.56%	1.34%	0.46%
"13-18"	1.99%	1.62%	-2.02%	0.32%	2.44%	3.17%	1.83%

Table 3: Modal Split in the EU, Source:

from the original work and some numerical results might be misleading.

1. the fleet of tankers and gas carriers is relatively old so fresh capital is required; nevertheless, it is not clear if the investment is justified given the swift to fuel mixes other than oil.

2. general cargo ships are leaving the marker or replaced by multipurpose (MPP) ones; therefore the ratios in the last column are very high as they disregard the revenues from recycling.

The results of Table 3 highlight another aspect of the problem, as they newbuliding column implies the financial benchmark for a new ship in every segment. As these results are based on aggregate global statistics, the comparison with reported purchase price of newbuildings in given shipbuild-ing markets, such as China, South Korea and Japan, should indicate premiums of size, delivery time and other features, such as speed and consumption. On the other hand, the retrofit column suggests the necessary investment per ship type in order to comply with the new environmental frame-work of regulations and offers a benchmark for the assess-ment of the decision of fleet renewal, selling or retrofitting the vessel. However, these results do not significantly vary from the reported costs in Table 2. By considering the rev-enues of demolition in the last column, the segment of pas-senger roro ships should attract the interest of financiers and technology providers.

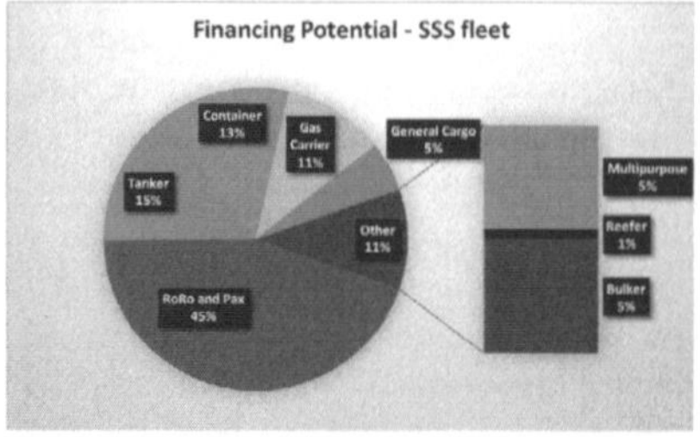

Figure 4: Breakdown of Estimated Financial Significance per Segment of SSS Fleet

3. ASSESSMENT OF THE MARKET POTENTIAL FOR WASP

The results of Section 3 are further processed and Tables 4 and 4 are derived. The total sums of Table 3 indicate the-oretically the total market for greening technology and re-lated retrofits. Assuming a parameterization of 10% to 50%for the WASP potential, i.e. assuming that WASP technolo-gies will tap only a part of the market, a market size of min USD1.1bn to a max USD5.6bn is estimated. Own as-sessment of 20-30% market shares for tankers and bulkers, 10-20% for roro-pax and MPP, and 10% for all other seg-ments yield a market potential of min USD5.9bn and max USD8.5bn. This assumption is rather conservative and re-flects also some technical considerations when considering the installation of WASP, namely [19]:

	# of vessels	NB		retrofit		ttl	ttl market
Bulker	1'948	632'336'124	28%	1'664'797'850	72%	2'297'133'974	1.45%
Tanker	10'475	3'010'500'482	51%	2'873'124'897	49%	5'883'625'378	4.76%
Container	2'498	4'330'115'627	78%	1'221'299'490	22%	5'551'415'117	8.10%
Gas Carrier	1'278	2'813'999'765	80%	684'251'942	20%	3'498'251'706	5.37%
General Cargo	15'539	2'530'056'621	60%	1'669'671'780	40%	4'199'728'401	106.64%
Multipurpose	3'106	1'471'728'169	53%	1'295'015'761	47%	2'766'743'930	55.38%
RoRo and Pax	10'435	15'145'851'628	90%	1'721'403'272	10%	16'867'254'900	94.03%
Reefer	1'468	1'062'468'719	91%	107'024'670	9%	1'169'493'389	99.09%
Grand Total	46'747	30'997'057'135	73%	11'236'589'660	27%	42'233'646'795	

Table 4: SSS financial needs, Source: own elaboration of published data, [17]

	# of vessels	NB	retrofit	significance incl. scrap
Bulker	1'948	7'527'811	4'278'931	5.28%
Tanker	10'475	5'648'219	2'007'087	15.14%
Container	2'498	16'527'159	3'191'275	13.50%
Gas Carrier	1'278	22'693'546	3'021'031	10.86%
General Cargo	15'539	5'548'370	1'135'701	5.31%
Multipurpose	3'106	14'288'623	3'297'576	4.79%
RoRo and Pax	10'435	18'606'697	1'638'694	44.60%
Reefer	1'468	27'242'788	1'796'843	0.52%

Table 5: Estimation of financial needs per vessel, Source: own elaboration

- availability of deck space;
- stability;
- local strength;
- local protection, e.g insulation, painting, corrosion;
- cabling for operation and control;
- required energy;
- obstruction of the visibility from the bridge;
- obstruction of cargo handling operations, e.g. interference with the operation of cranes and hatches.

As in the previous case the assessment of the market per number of vessels, Table 4 should be critically and carefully assessed. It reflects a reasonable and averaged number of ships per segments, that could support market estimations for WASP technologies promotion.

4. CONCLUSION

The aim of this work is to approximate the ship finance needs for SSS vessels, and estimate also the market potential for retrofitting with greening technologies and in particular of WASP. Hence, this paper aims to address the gap in the literature on the market potential for SSS vessels as well as for the implementation of WASP technologies in the SSS fleet.

Methodologically, this work is based on a recently published article, that estimates the financial gap for the greening of the global fleet. It is a bottom-to-top approach that estimates short-term needs (2025-6), considering the limitations when estimating the segments of the global fleet, as well as the average prices of assets, demolition and retrofits reported. In this paper, data of the SSS fleet, i.e. > 30kDWT, from the global fleet are extracted and further elaborated.

As key takeaways from this work, one should consider the following:

1. the total market for ship finance in SSS amounts USD42bn, i.e. circa 12% of the total exposure of the top-40 banks;
2. as expected circa 70% of this sum should be directed to fleet renewal and 30% to retrofit of the existing vessels;
3. given that WASP will tap only a part of the total retrofit market, the short-term market size ranges from USD5.9bn to USD8.5bn;
4. tanker, containerships and roro-passenger vessels seem to offer higher potential than other segments when con-sidering retrofitting.

To the best knowledge of the author, there is no similar work or report available to compare the above estimations, hence this work is rather novel. Nevertheless, this work is faced with many limitations, namely:

1. the cut-off threshold of > 30kDWT should be revisited per segment;
2. the estimation of the cost per segment for both new-builds and retrofits might be further refined, as there is no linear relationship of cost and performance; it is re-minded that marine equipment investments as well as newbuilding projects are lumpy, thus deeming averag-ing a challenge.

	retrofit	10%	20%	30%	40%	50%
Bulker	1'664'797'850	166'479'785	332'959'570	499'439'355	665'919'140	832'398'925
Tanker	2'873'124'897	287'312'490	574'624'979	861'937'469	1'149'249'959	1'436'562'448
Container	1'221'299'490	122'129'949	244'259'898	366'389'847	488'519'796	610'649'745
Gas Carrier	684'251'942	68'425'194	136'850'388	205'275'582	273'700'777	342'125'971
General Cargo	1'669'671'780	166'967'178	333'934'356	500'901'534	667'868'712	834'835'890
Multipurpose	1'295'015'761	129'501'576	259'003'152	388'504'728	518'006'304	647'507'881
RoRo and Pax	1'721'403'272	172'140'327	344'280'654	516'420'982	688'561'309	860'701'636
Reefer	107'024'670	10'702'467	21'404'934	32'107'401	42'809'868	53'512'335
Grand Total	11'236'589'660	1'123'658'966	2'247'317'932	3'370'976'898	4'494'635'864	5'618'294'830

Table 6: Assessment of the market potential for WASP in USD

	# of vessels	10%	20%	30%	40%	50%
Bulker	1'948	195	390	584	779	974
Tanker	10'475	1'048	2'095	3'143	4'190	5'238
Container	2'498	250	500	749	999	1'249
Gas Carrier	1'278	128	256	383	511	639
General Cargo	15'539	1'554	3'108	4'662	6'216	7'770
Multipurpose	3'106	311	621	932	1'242	1'553
RoRo and Pax	10'435	1'044	2'087	3'131	4'174	5'218
Reefer	1'468	147	294	440	587	734

Table 7: Assessment of the market potential for WASP in Number of Ships

3. a clustering of SSS operating in stricter regional sea areas, such as the emission control areas of Europe and of the US, could improve the overall approximation, and

4. a calibration of the results with unit cost of retrofit or newbuilds (USD/DWT) could also benefit the overal estimation, while the correlation of this unit cost with the energy performance improvement, could also sup-port further decarbonization benchmarking.

Given the importance of SSS, especially in the EU, the es-timation of the financial need offers a solid basis for the as-sessment of feasibility and viability of some policy objec-tives. As the top-40 banks retreat, then gearing of own cap-ital is required, and this implies also sufficient profit mar-gins, in particular adequate freight rates that can support rel-evant decision-making. Alternative sources of finance are available with a higher cost, that usually deems most project unattractive for further financing. Therefore, the environ-mental benefits of SSS should be internalized and reflected in the investment appraisal, or other policies should be de-ployed in order to accelerate and support further the decar-bonization of shipping. Further research in this direction can enlighten all stakeholders, such as operators, investors, and policy makers, especially in internalizing environmental benefit or by supporting shared economy models [see, e.g., 18], that promote leasing of equipment under shared risks and benefits schemes.

5. REFERENCES

[1] R. Aronietis, C. Sys, and T. Vaneslander. Ship retrofit solutions: economic, energy and environmental im-pacts. In S. Ehlers, editor, *Conference on Maritime-Port Technology (MTEC)*, Trondheim, Norway., Oct 27-29 2014.

[2] T. Chou, V. Kosmas, K. Renken, and M. Acciaro. New wind propulsion technology: A literature review of recent adoptions. Technical report, WASP Project, 2020. URL https://vb.northsearegion.eu/public/files/repository/20201013105303_WASP-WP4.D5B_NewWPT-ALiteratureReviewofRecentAdoptions-Final. pdf.

[3] DNV-GL. Retrofitting: Where the savings are, 2015. URL https://www.dnvgl.com/maritime/energy-efficiency/article/retrofitting.html.

[4] European Commission. The future development of the common transport policy - a global to the construction of a community framework for sustainable mobility. Communication from the Commission to the European Parliament and the Council COM (1992) 494 final, Eu-ropean Commission, 1992.

[5] EUROSTAT. Transport – overview. URL https://ec.europa.eu/eurostat/web/transport.

[6] J. Faber, B. Behrends, and D. Nelissen. Analysis of ghg marginal abatement cost curves. Technical report, CE Delft, 2011. URL https://www.cedelft.eu/publicatie/analysis_of_ghg_marginal_abatement_cost_curves/1155.

[7] IMO. 2014 Guidelines on the Method of Calculation of the Attained Energy Efficiency DesignIndex (EEDI) for New Ships. *MEPC.245(66)*, 2014.

[8] IMO. 2014 Guidelines on the Method of Calculation of the Attained Energy Efficiency DesignIndex (EEDI) for New Ships. *MEPC.254(67)*, 2014.

[9] IMO. Prevention of air pollution from ships. Press Release, 2018.

[10] J. Ling-Chin and A. Roskilly. Investigating a conven-tional and retrofit power plant on-board a roll-on/roll-off cargo ship from a sustainability perspective – a life cycle assessment case study. *Energy Conversion and Management*, 117:305 – 318, 2016. ISSN 0196-8904. doi: https://doi.org/10.1016/j.enconman.2016.03.032. URL http://www.sciencedirect.com/science/article/pii/S0196890416301637.

[11] R. Lu and J. W. Ringsberg. Ship energy perfor-mance study of three wind-assisted ship propulsion technologies including a parametric study of the flet-tner rotor technology. *Ships and Offshore Struc-tures*, 15(3):249–258, 2020. doi: 10.1080/17445302. 2019.1612544. URL https://doi.org/10.1080/ 17445302.2019.1612544.

[12] H. Psaraftis, editor. *Green Transportation Logistics; The Quest for Win-Win Solutions*, volume 226 of *In-ternational Series in Operations Research & Manage-ment Science*. Springer International Publishing, 1st edition, 2016.

[13] H. N. Psaraftis and O. D. Schinas. Concerted action on shortsea shipping: State of the art study part 1. Techni-cal report, European Comission - DG Transport, July 1996.

[14] H. Ross and O. Schinas. Empirical evidence of the interplay of energy performance and the value of ships. *Ocean Engineering*, in print, 2019.

[15] O. Schinas, editor. *Reduction of GHG Emis-sions from Ships; Wind propulsion solutions*, num-ber *MEPC75/INF*.26, November 2020. International Maritime Organization (IMO). doi: $10.13140/RG.2. 2.18835.02081$.

[16] O. Schinas. *The Impact of Air Emissions Regulations on Terminals*, pages 213–244. Springer International Publishing, 2020. ISBN 978-3-030-39990-0. doi: $10. 1007/978-3-030-39990-0 10$. URL $https://doi.org/10.1007/978-3-030-39990-0_10$.

[17] O. Schinas and N. Bergmann. The short-term cost of greening the global fleet. *Sustainability*, 13(16), 2021. ISSN 2071-1050. doi: 10.3390/su13169439. URL https://www.mdpi.com/2071-1050/13/16/ 9439.

[18] O. Schinas and D. Metzger. A pay-as-you-save model for the promotion of greening technologies in ship-ping. *Transportation Research Part D: Transport and Environment*, 69, 2019.

[19] O. Schinas and D. Metzger. Financing ships with wind-assisted propulsion technologies. In *RINA Wind Propulsion, 15th -16th October, London*. Royal Insti-tution of Naval Architects, 2019.

[20] O. Schinas and H. Psaraftis. New frontiers through shortsea shipping. In *SNAME Annual Conference Pro-ceedings*. SNAME, 1997.

[21] Shiipping Intelligence Network - Clarksons'. Sin clarksons', August 2011. URL http://sin.clarksons.net/.

[22] V. Stulgis, T. Smith, N. Rehmatulla, J. Pow-ers, and J. Hoppe. Hidden treasure: Financial models for retrofits. Technical report, Carbon War Room - Research Report, 2014. URL http://www.carbonwarroom.com/sites/default/files/reports/CWR%20Shipping%20Efficiency%20Finance%20Report.pdf.

[23] J. von Wirén. Norsepower rotor sails confirmed savings of 8.2 co2 in maersk pelican project, 2019. URL https://maersktankers.com/media/norsepower-rotor-sails-confirmed-savings.

6. AUTHORS' BIOGRAPHIES

Orestis Schinas is founding partner of the HHX.blue, a spe-cialised maritime fin-tech entity that focusses on greening, as well as Professor of Shipping & Ship Finance at the Ham-burg School of Business Administration (HSBA). His pro-fessional career includes large corporate projects, such as IPO and advanced research and business development as-signments in the finance, shipping and port sectors and has provided his services to private and public entities in Eu-rope, Africa and the GCC, as well as to international organi-zations. His work focuses on the commercial and financial aspects of maritime ventures while he retains a keen interest on sustainable finance and business development.

ACCELERATING THE UPTAKE OF WIND PROPULSION IN SHIPPING

Z Rafaelova, L Pomaska, V Kosmas and **M Acciaro,** Kühne Logistics University, Germany

SUMMARY

Wind Assisted Ship Propulsion (WASP) is seen as a potential solution to the environmental challenges of the maritime industry, still broader uptake is limited by barriers that hinder its adoption, implementation and upscaling. This study identifies overcoming strategies to accelerate the uptake of WASP by means of a systematic literature review. The study suggests that introducing and applying open innovation strategies in the commercialization phase lead to valuable results. In particular, proactive environmental strategy (PES), cooperation strategy, innovation activities, financing and policymakers were found to be vital key concepts for the implementation of new innovation technologies. The findings provide managerial and policy insights for all stakeholders that are involved in the process, mainly WASP technology providers, shipowners and policymakers.

1. INTRODUCTION

The maritime industry is under pressure to mitigate its environmental impact and deliver changes in line with the limits laid out in the Paris Climate Agreement [1] by implementing strict emission regulations. Full decarbonization of international shipping is necessary and highlights the importance of short- to medium-term emission reduction measures to meet this goal [2]. As a result of the decarbonization efforts, various options and the uptake of greener technologies are currently investigated. Wind Assisted Ship Propulsion (WASP) technologies are seen as an effective and innovative option that can benefit shipowners and society through fuel consumption reduction, energy efficiency improvement and emissions reduction [3]. Despite the growing attention and interest of key industry stakeholders due its promising fuel savings and environmental benefits, WASP technologies have not yet reached market maturity [4]. While many studies have investigated advantages and benefits of WASP technologies [5], [6] there is no specific study that has identified strategies for overcoming barriers and hence accelerating WASP adaption, implementation and upscaling. This paper therefore addresses this gap through an assessment of barriers that are preventing the uptake of WASP technologies and proposing strategies found outside of the maritime literature that have been developed and applied for the implementation of greening innovations. In this context, it is important to investigate following research question: Which strategies can be applied from industries outside shipping for overcoming the WASP implementation barriers?

In order to answer the research question, the authors follow a systematic literature review approach. This review paper identifies strategies that can be adapted in order to overcome the preventing factors. The purpose of this paper is to create a framework that shipowners can consult or adapt to their specific needs in order to evaluate strategies for investments in CO_2 abatement technologies. It creates a foundation of suggestions for WASP technology providers that can be adapted internally to overcome certain market barriers. It also provides a basis for policymakers with respect to regulation and incentives that can be provided to accelerate the uptake of WASP technologies.

The paper proceeds with the following structure. Section 2 highlights the barriers that prevent the uptake of WASP in the shipping industry. Section 3 presents the theoretical framework of this paper and the methods used within this study namely systematic literature approach. Section 4 presents key strategies from the literature review that are applied in other industries. The strategies are then put into a WASP context. Section 5 concludes and provides recommendations for future research.

This is a revised and redacted version of research that has appeared in [7] as well as work carried out in the context of the Interreg North Sea Region Project "WASP".

2. BARRIERS OF THE UPTAKE OF WASP

According to the literature [3], [5], [6], [8] the installation of wind technologies on ships provides numerous benefits, for instance lower carbon emissions and reduction of the exposure to the price volatility of fossil fuels [9]. Still, there are factors that are preventing the adoption of WASP [4].

2.1 COST AND ACCESS TO CAPITAL

There is an existing financial burden in access to capital for WASP providers, due to the small size of the market [4]. On the demand side, it is problematic for shipowners to install WASP technologies caused by the large initial investment, lack of funding for conducting trials and overall lack of liquidity in the sector. The conventional financing schemes appear to be reluctant to providing financial support, owing to the inability for shipowners to fulfil their contribution, as the payback period of these technologies is not fixed [4]. Hidden costs are described as non-market failures which occur after the implementation of the technology namely maintenance and costs of spare parts and operation. The more complex the technology is, the less attractive it is for the shipowners, due to the risks of increasing the burden on the crew [10]. Other risks include the fleet, market, and operational heterogeneity of the shipping industry [11]. WASP technologies are proven to be cost-effective on

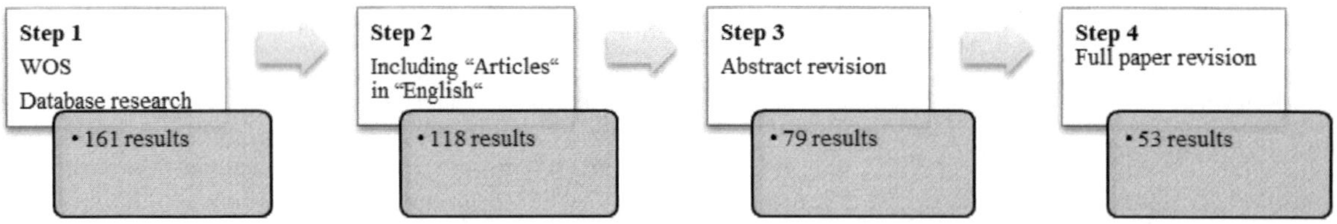

Figure 1 Searching and selecting process. Source: Authors

average, but there are not many evidences on all specific cases of WASP performance, as it varies based on ship sizes, type and route [12]. For instance, the technology's payback period could be lengthened if operated in non-favorable with less windy regions [11]. Subsequently, fuel prices are observed as a structural driver of the economic barrier. Other factors are the role of institutions and interaction among technology providers. Rojon and Diepernick [13] argue that some of the barriers in accessing capital are created by the lack of policies in promoting wind propulsion, public finance of R&D activities, start-up finance, weak compliance and enforcement mechanisms. Furthermore, the lack of knowledge exchange incentives and platforms among technology providers is preventing from lobbying activities that can influence policymakers [13].

2.2 SPLIT INCENTIVES

The legal setup of the shipping industry is one aspect that creates split incentives for implementing wind technologies [4]. According to Acciaro et al. [10], charterers are perceived as a bigger barrier to the implementation of new technologies in general, outweighing the role of regulations. Depending on the type of agreement, charterers are profiting from fuel expenditure savings most of the time, which creates no motivation for shipowners to take the risk of retrofits [4]. Additional barrier are the commercial agreements, as changes in routing or speed for the optimization of the WASP technology on board might not be allowed [11]. When examining the WASP target market in the face of dry bulk, it can be observed that 60% is under charter contracts. The duration of 90% of those contracts is less than two years, which is less than the payback period of the WASP technology [13]. Furthermore, the cost-efficiency of WASP technologies might be outperformed by simple cost-efficient solutions causing greater cost savings [4].

2.3 TRUSTED INFORMATION

One of the most prominent concerns of shipowners is related to the overall operability of WASP. For this reason, stakeholders require detailed, trusted information on their specific case, that include all aspects such as setup, maintenance, opportunity costs, compatibility with other technologies and infrastructural limitations [4]. However, availability is not enough as the source of information plays a crucial role and data from technology providers is generally perceived as biased and therefore not trusted [13]. Hence, lack of trust among actors and the overall conservative and risk-averse character of the industry are preventing the exchange of information, especially among technology providers due to the fear of losing competitive advantage on the market [13]. Another obstacle is derived from the variability of performance as savings from specific cases cannot be generalized. It is necessary to factor in operating conditions such as wind speed, wave height, seasonal patterns, route, crew training, trade patterns, trip duration, trip irregularity and port calls [3]. Subsequently, stakeholders should be provided with case specific parameters of a ship and its routings. This faces another hurdle since there is no standardized way of collecting this data. Currently, the available information is based on a small-scale theoretical tests, due to the limited number of funds. Another reason is the skepticism on the side of shipowners, that are usually opposed to the idea to participate in trails, due to the lack of legal protection from arising risks [4].

3. RESEARCH DESIGN

For the purpose of investigating and presenting potential strategies in implementing technological innovations, a systematic literature approach was applied, to allow for using a large and diverse set of papers. For the collection of relevant materials for revision, Web of Science Core Collection was used, as it is ranked as one of the most widespread literature databases [14] and allows for a focused search of papers [15]. Figure 1 displays the searching and the selecting process that were performed. As a first step, key search terms were identified, following the example by Pittaway et al. [16], using a brainstorming technique. This activity resulted in identifying strategies, technologies, innovation, and barriers as leading topics to be searched for. Based on these terms, a search string was defined as an AND function, consisting of the combination of strateg*, technolog*, innov*, barrier* terms. This type of search insures the relevance of the results and inclusion of only relevant materials. The search was narrowed down to title and abstract to ensure that those terms are included in the main topic of the paper, which gave 161 results. To narrow down the results, a selection process was performed using following filters: exclusion of proceeding papers, editorial material and notes; exclusion of articles that were not in English; exclusion of articles that were published out of the time span of 2005-2021. At this point, 118 articles were obtained but they needed to be analyzed more precisely. A revision process was conducted to ensure quality assessment of the results. First, the obtained articles were reviewed by title and abstract. As a result, 39 articles were excluded. The most common exclusion reason was based on the fact that articles have presented strong characteristics of industries such as business to consumer or service-related context which are not relevant to the

context of WASP. The obtained 79 articles went through full text revision as a final step. The papers that did not provide context-relevant insights or did not present any strategies were excluded. The final result contains 53 papers.

4. RESULTS

4.1 OPEN INNOVATION

Open innovation (OI), which is described as the way to use knowledge inflows to accelerate internal innovation and knowledge outflows to stimulate external use of innovation and market growth [17], was fully or partially present in all articles. Currently, firms pursue a more closed approach to innovation [18], while OI is driven by pioneering companies that tend to approach the innovation process systematically and dedicate a corporate venturing unit for external technology exploitation activities [19]. In the near future, external technology acquisition and exploitation will be a necessity rather than an option, implying that a closed innovation approach has negative effects on the long-term performance of the company. Thus, proactive openness to innovation holds competitive advantages for innovators, independent from the industry context [18]. *Proactive environmental strategy (PES), cooperation strategy, innovation activities, financing and policymakers' role* are crucial topics in examining this strategy, which will be described further.

4.1 (a) Proactive environmental strategy (PES)

PES has a significant mediating role in the relationships between interorganizational cooperation, knowledge sharing, and technological eco-innovation. PES constitutes a unique organizational capability that affects company's environmental and economic performance through improved transformation of knowledge. In essence, PES is in place, when companies not only take actions just enough to meet the minimum requirements by the regulators, but voluntarily engage in activities to reduce their environmental impact [20]. The success of PES lies in the introduction of coherent activities across multiple departments of the company [21]. Additionally, improved transformation of knowledge is crucial [20], thus creating a *knowledge sharing* setup shows the stimulation of the overall innovation ecosystem and individual actor's performance [19]. In general, knowledge sharing can be internal or external. One aspect of external knowledge sharing is forming alliances [22] as it holds benefits such as stable cooperation between collaborative agents [23]. The establishment of perfect information sharing mechanism, to develop a variety of open communication channels is suggested in order to accelerate co-contribution and keeping conflicts within alliances on a moderate level [24]. Another aspect of external knowledge sharing is *internationalization* and therefore establishing international relationships for a company. International support programs are perceived as a key to establishing innovativeness and competitiveness in the case of small and medium enterprises (SMEs) targeting international markets. For companies that are focusing primarily on domestic markets, support on client's acquisition is prevailing [25].

A significant part of knowledge sharing is the *communication* between actors for the success of industrial technological innovation, which can be observed in three different aspects, namely on institutional, market and company level [26]. On industrial level, deeper exchange and cooperation among firms should be carried. On market level, there are three major directions for improvement, namely market-oriented development and resource allocation, increasing applied research to stimulate innovative research and lastly, having scientific research institutions as supervisors. On a company level, communication barriers can be approached through information technology (IT) improvements [26].

4.1 (b) Cooperation strategy

The implementation of cooperative strategies is suggested for smaller sized markets with fewer competitors [19]. SME's can successfully collaborate with external firms, by using a blend of strategies on multi-firm collaboration modes in the commercialization phase, namely networks with a lead partner, equal partnerships, and partnership for external technology commercialization [27]. Companies interact and exchange good practices and knowledge with partners in the face of research centers and universities within a co-creation process [28]. Networking strategy is another tool for building collaborative capabilities within the organization and between external partners [29]. Especially external cooperation is considered vital due to enhanced IP management capabilities with various external sources [30]. Furthermore, partnerships with complementary products are beneficial and a stepwise diffusion strategy suggests starting a pilot test of both technologies and business models in one country only. Additionally, a more consumer centric approach is suggested where customers are involved in the early stage of the product development process to understand customer needs [31]. Moreover, product and process quality can be improved by cooperating with external researchers that have industry independent judgements. For instance, university business incubators (UBI) aim at stimulating the formation of new ventures by providing mentoring and administrative support for firms [19].

Another aspect is the sourcing of simplified high-end technologies through research institutions or universities by using inputs from SME to capture market needs [32]. The simplification strategy is in line with the trend of OI and can be achieved on technical as well as on market level through the value proposition that the product holds [33]. The protection of product technology is crucial whereas licensing is considered as an effective way for technology acquisition and utilization in innovation technology industries [34].

4.1 (c) Innovation activities

Strategic orientations are defined as a tool for achieving competitive advantage, based on the specific firm's context [35], [36]. There are two major differentiation strategies namely *market orientation* and *technology orientation* [31], where the former type is based on sustaining product innovation, and the latter type stresses market and product image management [37]. Market-oriented companies have relatively higher sales performances, whereas technology-oriented score better in company value. However, the combination of both does not bring outstanding results, thus, in order to achieve the desired outcomes in context of innovative companies, managers should set their focus on one prevailing orientation only [38].

In terms of *technology orientation*, long-term growth can be achieved through the created high value among innovative firms [38], where technological capabilities foster innovation exploitation at an accelerated rate and strategic flexibility strengthens the positive effect [24]. It is suggested to develop market entry strategies and to define a strategic roadmap with goals, objectives and the focus of R&D areas [39]. Moreover, the implementation of technology acquisition strategy has an influence on technology innovation performance, along with the combination of technology innovation strategy (TIS) and Intellectual Capital (IC) development [40]. The complementary relationship between internal and external innovation, which in combination with internal and external knowledge strategy creates innovation activity, has a positive effect on marginal return on all individual activities [41]. Subsequently, attention should be paid equally to organizational, technical and innovation resources [42].

On the contrary, *market orientation* strategy shows a positive effect on innovation results by focusing on commercialization activities and the combination of both internal and external R&D activities [31]. Marketing strategy innovativeness (MSI) refers to the degree of differentiation between the marketing strategy for a particular product, compared to the convectional strategies applied for this case. It should be bundled with external managerial relationships and market dynamism through new distribution channels, media content or payment methods [31]. It is essential for companies to keep their launch approaches flexible to align with the continuous market demand [43]. Furthermore, marketing promotion strategy that emphasizes on promoting benefits of sustainable production combined with the launching strategy stimulates the demand across customers [29]. Thus, OI strategies are efficiently applied only if introduced in the commercialization phase, rather than the research and development phase [44].

4.1 (d) Financing activities

Finance-related services are vital for new technology-based firms to persuade a technology strategy [44]. Environmental costs should be incorporated in the planning and investment incentives in combination with delayed implementation. Miller and Hope [45] suggest the establishment of proper feed-in tariffs and geographical diversification of installation and capacity building in commercial banks. Another alternative is the introduction of a subsidy system of high-tech guarantee loss compensation [46]. Furthermore, product as a service is a viable strategy, where companies can monitor, measure, customize and bill for assets use in a great detail. This allows for purchasing units of a product and accounting for them as a variable cost, rather than capital investment. Finally, the exploitation of public funds is suggested as a way to prove higher standards of the offered product [28].

4.1 (e) Policymakers' role

The role of institutions is crucial for companies when trying to overcome industry-related barriers as they have control over knowledge production and diffusion, alignment of actors, reduction of uncertainty (by including risk reduction) and control over knowledge usage [47]. Policymakers can be involved by introducing tax-braking for research activities in order to positively influence return on investments and to stimulate technology sourcing [22]. Moreover, knowledge sharing, innovation performance improvement and financial support at different diffusion stages enhances the enabling environment of innovation technology industries [48]. Economic incentives should be provided to ensure participation of various actors in the technology diffusion stage [49]. Investment schemes such as co-investment, loans or risk-guarantees and partnerships between public and private sector for leveraging the interest for both parties are instruments of those incentives [50]. Additionally, international financing through funds, grants and international funding projects can be eligible. Formal lobbying through workshops and support by local government and informal lobbying through campaigns for the public are other tools for policymakers [50]. Furthermore, governments can take decisions toward specific goals on technology acquisition, development and deployment to improve the domestic technology with a technology leading approach, which is called government technology policy [51]. Enterprise technology strategy formulates the execution plan for exploiting technological resources and for achieving improved performance [51]. Finally, regulatory and environmental strategy is a tool for the formulation of policies to promote sustainability practices in the context of manufacturing organizations' ecosystem, for instance, the support for developing IP competences for green products and process innovation [29].

4.2 DISCUSSION

Potential strategies that are suggested from the managerial literature are put into maritime context in order to overcome implementation barriers for WASP. OI plays a central role in strategies for technological innovation uptake and can be adopted by shipowners and technology providers in the context of WASP [18]. Various suggestions for policymakers are provided to stimulate the process of OI between the different actors on the market and to provide a better setting for accomplishing sustainability-oriented goals.

4.2 (a) Strategies for cost and access to capital barrier

On the supply side, stimulation of innovation activities for technology providers leads to company growth and strengthens the industry. Through the means of internationalization, technology providers develop their own capacities and enhance their competitive advantages. The supply to shipowners is improved by enhancing competition within technology providers' markets which results in overall market growth due to the attractiveness and potential for new entrepreneurs. This leads to enhanced investment opportunities due to the enlarging size of the market and allows for significant portfolios of the technology performance use cases. Another way for technology providers to ensure the coverage of more use cases is cooperation between one another by applying multi-firm collaboration modes in the commercialization phase, namely networks with a lead partner, equal partnerships, and partnership for external technology commercialization. A clear governmental policy and delayed implementation strategy are beneficial to overcome high initial investment cost of WASP technologies and the lack of assurance in supply. Additionally, product as a service strategy enables technology providers an innovative way of selling products by charging for maintenance or specific parts only. Hence, shipowners have the opportunity to account for them as variable costs, rather than capital investment. Governmental support in terms of economic incentives can be provided through program-based crediting mechanisms, investment schemes, loans or risk-guarantees and partnership between public and private sectors for leveraging interest for both parties. For overcoming long return on investment barriers, policymakers support with incentives for instance tax-breaking for research activities. Finally, specific funds for financing wind propulsion projects, that are supervised by qualified personnel for effective distribution, are essential.

4.2 (b) Strategies for split incentive barrier

The role of policymakers is crucial to overcome split incentive barriers, as they possess the power to create incentives for the implementation of technological innovation. Promoting sustainability practices in WASP development ecosystems with the means of regulatory and environmental strategies is vital for overcoming institutional barriers, such as support for development of IP competences in green and sustainable context. In the case of no monetary benefits that could have been provided by policymakers, economic and inventive-based strategy aims at allocating separate funds for sustainability innovations to stimulate a mechanism for creating an incentive to implement WASP.

4.2 (c) Strategies for information trust barrier

One strategy that was suggested is the launch strategy which needs to be consistent with the marketing promotion strategy in promoting benefits of sustainable products and practices. This helps creating trust among shipowners and investors and stimulates the demand for WASP. Shipowners heavily rely on quantitative methods for evaluating new opportunities, therefore information about the technology requires to be presented in transparent way by introducing environmental management systems for monitoring processes in the technology provider's company. Moreover, the inclusion of independent partners in the development phase ensures a better trusted quality and objective demonstration of the technology's usability. Subsequently, WASP providers can increase their competitive advantage through public programs participation as their performance will be compliant with certain standards and thus, more trustworthy. Another approach is the inclusion of research institutions within the use of the simplification strategy where WASP are simplified by using inputs from shipowners in order to capture better market needs. Furthermore, firm's cooperation through market-oriented development is advantageous for all stakeholders on the market and prevents competition among WASP providers. For high-tech markets such as the WASP, it is suggested to continuously improve testing of innovative achievements by shipowners with the supervision of scientific research institutions as verifiers. Additionally, knowledge sharing platforms for the interaction between market actors should be implemented. This helps in changing the shipowners passive role to an active role and therefore stimulates innovation. Knowledge sharing platforms in the case of WASP should be organized and managed by external organizations, for instance the EU, IMO or independent institutions. Lastly, governments can increase the pressure on shipowners to implement innovative technologies and at the same time improve functions with focus on knowledge diffusion, guidance of the search and market formation support.

5. CONCLUSIONS

In the introduction of this paper, the need for full decarbonization of the shipping sector through more sustainable technologies has been determined. WASP has been identified as an option to meet emission reduction targets, yet managerial and policy guidance for overcoming implementation barriers for this particular technology have not been researched in sufficient detail. The contribution of this paper is to address this gap by investigating strategies outside of shipping that have been suggested for the implementation of greening innovations in various industries. The approach of a systematic literature review is required in order to incorporate a significant and diverse set of papers and to present the literature in an integrated way.

The study stresses the need of prerequisites for implementing core principles suggested by OI strategy to overcome key barriers of cost and access to capital, split incentives and trusted information. In particular, PES, cooperation strategy, innovation activities, financing and policymakers were found to be vital key concepts for the implementation of new innovation technologies. The literature further suggests that introducing and applying OI strategies in the commercialization phase of SME leads to valuable results, which holds for WASP providers. The study found that it is advisable for policymakers to implement economic incentives through program-based crediting mechanisms, investment schemes and tax-breaking for research activities. The establishment of funds and international financing for wind propulsion projects help further accelerating the uptake of WASP. Partnerships between public and private sector leverages interests for both parties and enhanced political competition increases the optimal level of green technology innovation. Governmental policies should be developed to support IP competencies and therefore protect assets of technology providers. Furthermore, new policies should be implemented focusing on knowledge diffusion. This helps overcoming the information trust barrier by implementing knowledge sharing platforms that are organized by impartial international organizations. On market level, a first strategy implies close collaboration between WASP companies by forming alliances as it holds benefits of stable cooperation on both sides of the partnership. Environmental management systems for monitoring technology provider's performances is an effective tool to create trust for shipowners. Other strategies such as IP management, improvement of IT infrastructure on the market and simplification strategies tend to generate a positive effect in terms of perceived impact of innovation. In general, open innovation strategies are beneficial to implement in the adoption, implementation and upscaling of WASP technologies.

The study is limited to the selection of articles used within the systematic literature review based on the authors' definition of inclusion and exclusion criteria. Additionally, papers have been extracted from one database of high quality, however, other databases can complement this extracted foundation in future studies. Lastly, the suggested strategies could be tested through the means of surveys or interviews with industry experts, including stakeholders from the maritime industry for validation to further accelerate the uptake of WASP.

6. REFERENCES

[1] UNFCCC.INT, "United Nations Climate Change," United Nations, 2015. [Online]. Available: http://unfccc.int/paris_agreement/items/9485.php. [Accessed 11 August 2021].

[2] M. Traut, P. Gilbert, C. Walsh, A. Bows, A. Filippone, P. Stansby and R. Wood, "Propulsive power contribution of a kite and a Flettner rotor on selected shipping routes," *Appl. Energy*, vol. 113, pp. 362-372, 2014.

[3] T. Chou, V. Kosmas, M. Acciaro and K. Renken, "A Comeback of Wind Power in Shipping: An Economic and Operational Review on the Wind-Assisted Ship Propulsion Technology," *Sustainability*, vol. 13, no. 1880, pp. 1-16, 2021.

[4] D. Nelissen, M. Traut, J. Köhler, W. Mao, J. Faber and S. Ahdour, "Study on the analysis of market potentials and market barriers for wind propulsion technologies for ships," CE Delft, Delft, 2016.

[5] T. Smith, C. Raucci, S. Haji Hosseinloo, I. Rojon, J. Calleya, S. Suárez de la Fuente, P. Wu and K. Palmer, "CO2 emissions from international shipping. Possible reduction targets," UMAS, London, 2016.

[6] E.A. Bouman, E. Lindstad, A.I. Rialland and A.H. Stromman, "State-of-the-art technologies, measures, and potential for reducing GHG emissions from shipping - A review," *Transp. Res. D Environ.*, vol. 52, pp. 408-421, 2017.

[7] Z. Rafaelova, "Accelerating the uptake of Wind Propulsion in shipping," B.S Thesis, Kühne Logistics University, Hamburg, Germany, 2021.

[8] P. Balcombe, J. Brierley, C. Lewis, L. Skatvedt, J. Speirs, A. Hawkes and I. Staffel, "How to decarbonise international shipping: Options for fuels, technologies and policies," *Energy Convers. Manag.*, vol. 182, pp. 72-88, 2019.

[9] B. Snyder and M.J. Kaiser, "Ecological and economic cost-benefit analysis of offshore wind energy," *Renew. Energy*, vol. 34, pp. 1567-1578, 2009.

[10] M. Acciaro, P.N. Hoffmann and M.S. Ride, "The Energy Efficiency Gap in Maritime Transport," *J. Ship. Ocean Eng.*, vol. 3, no. 10, pp. 1-10, 2013.

[11] N. Rehmatulla, S. Parker, T. Smith and V. Stulgis, "Wind technologies: Opportunities and barriers to a low carbon shipping industry," *Mar. Policy*, pp. 217-226, 2017.

[12] S.C. Pirrong, "Contracting Practices in Bulk Shipping Markets: A Transactions Cost Explanation," *J. Law Econ.*, vol. 36, no. 2, 1993.

[13] I. Rojon and C. Dieperink, "Blowin' in the wind? Drivers and barriers for the uptake of wind propulsion in international shipping.," *Energy Policy*, vol. 67, pp. 394-402, 2014.

[14] A.N. Guz and J.J. Rushchitsky, "Scopus: A system for the evaluation of scientific journals," *Int. Appl. Mech.*, vol. 45, no. 351, 2009.

[15] M. Easterby-Smith, R. Thorpe and P. Jackson, *J* Management Research, Los Angeles: SAGE, 2021.

[16] L. Pittaway, M. Robertson, K. Munir, D. Denyer and A. Neely, "Networking and innovation: a systematic review of the evidence," *Int. J. Manag. Rev.*, vol. 5, no. 6, 2004.

[17] H. Chesbrough and A.K. Crowther, "Beyond high tech: early adopters of open innovation in other industries," *R&D Management*, vol. 36, no. 3, pp. 229-236, 2006.

[18] U. Lichtenthaler, "Open innovation in practice: An analysis of strategic approaches to technology transactions," *IEEE Trans. Eng. Manag.*, vol. 55, no. 1, pp. 148-157, 2008.

[19] A. Crupi, N. del Sarto, A. di Minin, R. Phaal and A. Piccaluga, "Open innovation environments as knowledge sharing enablers: the case of strategic technology and innovative management consortium," *J. Knowl. Manag.*, vol. 25, no. 5, pp. 1263-1286, 2021.

[20] A. Ryszko, "Interorganizational cooperation, knowledge sharing, and technological eco-innovation: The role of proactive environmental strategy – empirical evidence from Poland," *Pol. J. Environ. Stud.*, vol. 5, no. 2, pp. 753-764, 2016.

[21] M.S. Bhatia, "Green process innovation and operational performance. The role of proactive environment strategy, technological capabilities, and organisational learning," *Bus. Strateg. Environ.*, pp. 1-13, 2021.

[22] J. Zhao and P. de Pablos, "Chinese firms' outward direct investment: Technological innovation mechanisms, organizational modes, and improving strategies. Human Factors and Ergonomics," *Manufacturing*, vol. 20, no. 2, pp. 149-160, 2010.

[23] M. Zhao and C. Li, "Motivation mechanism prevents adverse selection in industrial technology innovation strategic alliance," *Discrete Dynamics in Nature*, 2014.

[24] K. Zhou and F. Wu, "Technological capability, strategic flexibility, and product innovation," *Strateg. Manag. J.*, vol. 31, no. 5, pp. 547-561, 2010.

[25] S. Mueller-Using, W. Urban and J. Wedemeier, "Internationalization of SMEs in the Baltic Sea Region: Barriers of cross-national collaboration considering regional innovation strategies for smart specialization," *Growth Change*, vol. 51, no. 4, pp. 1471-1490, 2020.

[26] T. Cui, H. Ye, H. Teo and J. Li, "Information technology and open innovation: A strategic alignment perspective," *Inf. Manag.*, vol. 52, no. 3, pp. 348-358, 2015.

[27] K. Henttonen and H. Lehtimäki, "Open innovation in SMEs: Collaboration modes and strategies for commercialization in technology-intensive companies in forestry industry," *Eur. J. Innov. Manag.*, vol. 20, no. 2, pp. 329-347, 2017.

[28] M. Zeleny, "High technology and barriers to innovation: From globalization to re-localization," *Int. J. Inf. Technol. Decis. Mak.*, vol. 11, no. 2, pp. 441-456, 2012.

[29] H. Gupta, S. Husi-Sarpong and J. Rezaei, "Barriers and overcoming strategies to supply chain sustainability innovation," *Resour., Conserv., Recycl.*, vol. 161, 2020.

[30] S. Y. Kim and E. Kim, "How intellectual property management capability and network strategy affect open technological innovation in the Korean new information communications technology industry," *Sustainability*, vol. 10, no. 8, 2018.

[31] A.G. Frank, M.N. Cortimiglia, J.L.D. Ribeiro and L.S. Oliveira, "The effect of innovation activities on innovation outputs in the Brazilian industry: Market-orientation vs. technology-acquisition strategies," *Res. Policy*, vol. 45, no. 3, pp. 577-592, 2016.

[32] D. Yu and C. Hang, "Creating technology candidates for disruptive innovation: Generally applicable R & D strategies," *Technovation*, vol. 31, no. 8, pp. 401-410, 2011.

[33] C.M. Christensen and M. Raynor, The Innovator's Solution, Boston: Business School Press, 2003.

[34] M. Holgersson and O. Granstrand, "Patenting motives, technology strategies, and open innovation," *Manag. Decis.*, vol. 55, no. 6, pp. 1265-1284, 2017.

[35] G.S. Day, "The Capabilities of Market-Driven Organizations," *J. Mark.*, vol. 58, pp. 37-52, 1994.

[36] S.D. Hunt and M.M. Robert, "The Competitive Advantage: Theory of Competition," *J. Mark.*, vol. 59, pp. 1-15, 1995.

[37] X. Liu and X. Wu, "Technology embeddedness, innovation differentiation strategies and firm performance: Evidence from Chinese manufacturing firms," *Innov.: Manag. Policy Pract.*, vol. 13, no. 1, pp. 20-35, 2011.

[38] N.K. Park, U.D. Park and J. Lee, "Do the Performances of Innovative Firms Differ Depending on Market-oriented or Technology-oriented Strategies?," *Ind. Innov.*, vol. 19, no. 5, pp. 391-414, 2012.

[39] S.J. Edgett and R.G. Cooper, "Developing a Product Innovation and Technology Strategy for Business," *Res. Technol. Manag.*, vol. 53, pp. 33-40, 2010.

[40] C. Verbano and M. Crema, "Linking technology innovation strategy, intellectual capital and technology innovation performance in manufacturing SMEs," *Technol. Anal. Strateg. Manag.*, vol. 28, no. 5, pp. 524-540, 2016.

[41] N. Wang, M. Xiao and I. Savin, "Complementarity effect in the innovation strategy: internal R&D and acquisition of capital with embodied technology," *J. Technol. Transf.*, vol. 46, no. 2, pp. 459-482, 2021.

[42] J. Liu, A. Baskaran and S. Li, "Building Technological-Innovation-Based Strategic Capabilities at Firm Level in China: A Dynamic Resource-Based-View Case Study," *Ind. Innov.*, vol. 16, pp. 411-434, 2009.

[43] M.H. Hsieh and K.H. Tsai, "Technological capability, social capital and the launch strategy for innovative products," *Ind. Mark. Manag.*, vol. 36, no. 4, pp. 493-502, 2007.

[44] M. Knockaert, E. Vandenbroucke and A. Huyghe, "Unravelling the need for innovation support services in new technology-based firms: The impact of commercialization strategy," *Sci. Public Policy*, vol. 40, no. 1, pp. 85-96, 2013.

[45] D. Miller and C. Hope, "Learning to lend for grid solar power: policy lessons from World Bank loans to India, Indonesia, and Sri Lanka," *Energy Policy*, vol. 28, pp. 87-105, 2000.

[46] C.S. Wu, H. Hu and F.H. Chen, "Evaluating the enhancement and improvement of China's technology and financial services platform innovation strategy," *Springer Plus*, vol. 5, no. 1, pp. 3562-3573, 2016.

[47] S. Borrás, "System of innovation theory and the European Union," *Sci. Public Policy*, vol. 31, no. 6, 2004.

[48] M. Suzuki, "Identifying roles of international institutions in clean energy technology innovation and diffusion in the developing countries: Matching barriers with roles of the institutions," *J. Clean. Prod.*, vol. 98, pp. 229-240, 2015.

[49] P. Balachandra, K. Hippu Salk and R. Sudhakara, "Commercialization of sustainable energy technologies," Renewable Energy, vol. 35, no. 8, pp. 1842-1851, 2010.

[50] H.E. Edsand, "Identifying barriers to wind energy diffusion in Colombia: A function analysis of the technological innovation system and the wider context," *Technol. Soc.*, vol. 49, pp. 1-15, 2017.

[51] Y. Yang, J. Hong, G. Song and S. Hong, "Technology policy, technology strategy and innovation performance: Evidence from Chinese aircraft and spacecraft manufacturing," *Sci. Public Policy*, vol. 44, no. 5, pp. 620-630, 2017.

7. AUTHORS BIOGRAPHY

Zhanita Rafaelova is a Student Assistant at the Hapag-Lloyd Center for Shipping and Global Logistics (CSGL) of Kühne Logistics University (KLU) since 2018. Her role is focusing on promoting the work of the center, covering topics in the maritime transport such as environmental policies, green technologies and alternative fuels. She has been involved in the Wind Assisted Ship Propulsion (WASP) project, contributing research on strategies in overcoming the barriers wind propulsion is facing. She holds a BSc in Business Administration from Kühne Logistics University (KLU) and she is currently pursuing a MSc in Management at KLU.

Lara Pomaska is a Junior Researcher at the Hapag-Lloyd Center for Shipping and Global Logistics (CSGL) of Kühne Logistics University (KLU). Her research interests focus on sustainability within the shipping industry, particularly on alternative fuels and sustainable innovations. Moreover, she is investigating the economic impact of green technologies within the Wind Assisted Ship Propulsion (WASP) project and is responsible for the work package "Policy and viable business". She holds a MSc in Global Logistics and Supply Chain Management from the Kühne Logistics University.

Vasileios Kosmas is a Senior Researcher at the Hapag-Lloyd Center for Shipping and Global Logistics (CSGL) of Kühne Logistics University (KLU). His work focuses mainly on environmental policies, green technologies, and alternative fuels within the maritime transport industry. He has been involved in various projects related to sustainable shipping and he is the lead researcher of the "Policy and viable business" work package of the Wind Assisted Ship Propulsion (WASP) project. Dr. Kosmas is a member of the International Association of Maritime Economists. He holds a Ph.D. degree in Operations Management from Copenhagen Business School and a MSc in Maritime Transport with Management from Newcastle University.

Michele Acciaro is the Director of the Hapag-Lloyd Center for Shipping and Global Logistics (CSGL) and Associate Professor for Maritime Logistics at the Department of Operations and Technology at Kühne Logistics University (KLU), in Hamburg. His recent research work has focused on container transport, corporate sustainability in shipping and ports, and decarbonisation. He has been involved in several research projects on green shipping, and he is a member of the research team for the Interreg North Sea Region Project WASP and has collaborated extensively with the Clean Cargo Initiative. He is a member of the European Sustainable Shipping Forum and the European Sustainable Port Forum.

A REVIEW OF WIND-ASSISTED SHIP PROPULSION FOR SUSTAINABLE COMMERCIAL SHIPPING: LATEST DEVELOPMENTS AND FUTURE STAKES

L Khan, J J R Macklin, B C D Peck, O Morton, J-B R G Souppez, Aston University, UK

SUMMARY

With the current global warming crisis and contemporary concerns for sustainability, the transport industry is developing and implementing novel solutions to reduce greenhouse gases. With close to 90% of the world's goods relying on maritime transportation, responsible for 3% of global energy-related carbon dioxide (CO_2) emissions in 2019, there is a vital emphasis on reducing emissions. The latest legislation from the International Maritime Organisation has imposed even tougher sulphur oxide targets. On the other hand, emission intensity for CO_2 will need to be decreased by 70% in 2050, compared to 2008 figures. While operating measures and fuel alternatives are suitable in the short-term to meet these novel regulatory constraints, as the use of fossil fuels tapers off, the long-terms solution appears to reside in wind-assisted ships. Consequently, this study aims to identify viable solutions that could reduce emissions, focussing on three prominent technologies, namely sails, rotors and kites. Furthermore, this review provides guidance on the benefits and risks associated with each technology and recommends guidelines for performance prediction and associated constraints. Ultimately, future stakes in wind-assisted propulsion are highlighted, including the need for full-scale validation, the challenge in assessing environmental and economic impact, and the structural issues associated with wind-assisted propulsion systems.

NOMENCLATURE

ABS	American Bureau of Shipping
BV	Bureau Veritas
CFD	Computational Fluid Dynamics
Class NK	Nippon Kaiji Kyokai
CO_2	Carbon Dioxide
DNV GL	Det Norske Veritas Germanischer Lloyds
DoFs	Degrees of Freedom
EEDI	Energy Efficiency Design Index
GHGs	Greenhouse Gases
IMO	International Maritime Organization
ISO	International Organization for Standardization
LR	Lloyds Register
NOx	Nitrogen Oxides
NP	Norsepower
PPP	Performance Prediction Program
RANS	Reynolds-Averaged Navier Stokes
SOx	Sulphur Oxides
VPP	Velocity Prediction Program
WASP	Wind-Assisted Ship Propulsion

1. INTRODUCTION

Shipping and maritime transportation account for close to 90% of the world's goods transport [1], and was responsible for 3% of greenhouse gases (GHGs) emissions in 2019, a figure forecasted to grow to 15% by 2050 should no actions be taken [2]. Some forecasts are even most pessimistic [3], as depicted in Figure 1. The introduction of increasingly stringent international regulations related to nitrogen oxides (NOx) and sulphur oxides (SOx), as well as CO_2 [4], coupled energy efficiency design index (EEDI), aim to achieve more eco-friendly vessels.

There are however many limitations to the EEDI, including its applicability to new builds only, meaning the majority of the commercial fleet will not be covered until the 2040s [5]. The targets are also not deemed challenging enough, poorly accounting for the developments in electrical technologies and wind-assisted propulsion, and ultimately having only a small impact [6]. Nevertheless, ambitious regulations to achieve a sustainable shipping industry are a strong driver behind reduced emission vessels.

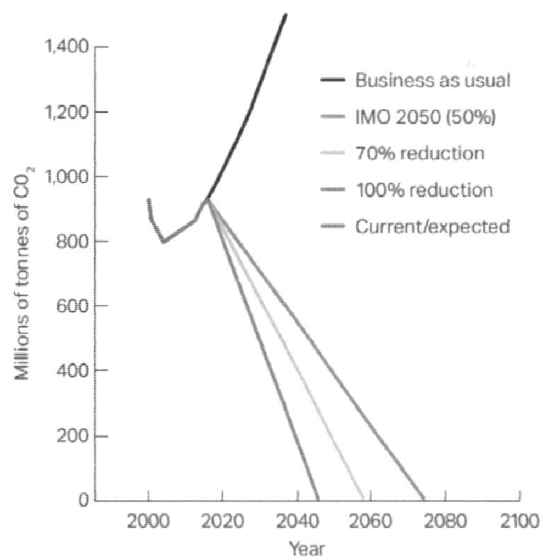

Figure 1: *CO_2 emissions forecast, taken from [3].*

To address the challenge of ship emissions, an array of strategies have been implemented. These can be categorized into operational and technological solutions. Operational measures [7] such as slow steaming (27% fuel saving for 10% speed reduction) and weather routing (2-5% GHGs reduction), while effective, will not be tackled in this paper.

Indeed, the focus is placed on technological advances to provide a long-term sustainable solution [8]. Amongst the various design technologies currently available, wind-assisted ship propulsion (WASP) shows the greatest potential to reduce GHGs, as presented in Table 1.

Table 1: *Effect of design technologies on GHG savings.*

Technology	Potential GHG Savings
WASP	In excess of 30% [7, 9, 10, 11]
Slender Design	Up to 15% [12]
Air Lubrication	Up to 13% [5, 7]
Increased Cargo	Up to 10% [7] (large vessels)
Materials	Up to 10 % [10]
Propeller design	Up to 10% [7]
Bulbous Bow	Up to 7% [6]
Heat Recovery	Up to 6% [14]
Hull Surface	Up to 5% [13]

This paper therefore addresses the current state-of-the-art in wind-assisted propulsion and related performance prediction in order to support the development of sustainable commercial shipping. The remainder of the paper is structured as follows. Section 2 introduces the background and main configurations for wind-assisted shipping. Section 3 tackles the principles and design considerations for sails, rotors and kites respectively. Then, Section 4 addresses the performance prediction. Future stakes are outlines in Section 5, with the key findings of this work summarized in Section 6.

2. WIND-ASSISTED SHIP PROPULSION

Amongst the various wind-assisted options available today, the six rig types originally investigated for wind-assisted ship propulsion in the 1980s [15], and pictured in Figure 2, remain relevant [5]. These are:

- Modern square rig (e.g. DynaRig),
- Rigid/Wing sails,
- Wind turbine,
- Soft sails,
- Rotating cylinders (including Flettner rotors and boundary layer suction devices),
- Kite.

It should be noted that the wind turbine approach is intended to produce electricity to then power the vessel [16], and as such will not fit within the scope of this paper, focused on wind-assisted propulsion options. Moreover, novel theoretical concepts, for instance the *Vindskip* 199 m carriers with an airfoil shaped hull [17], will not be discussed.

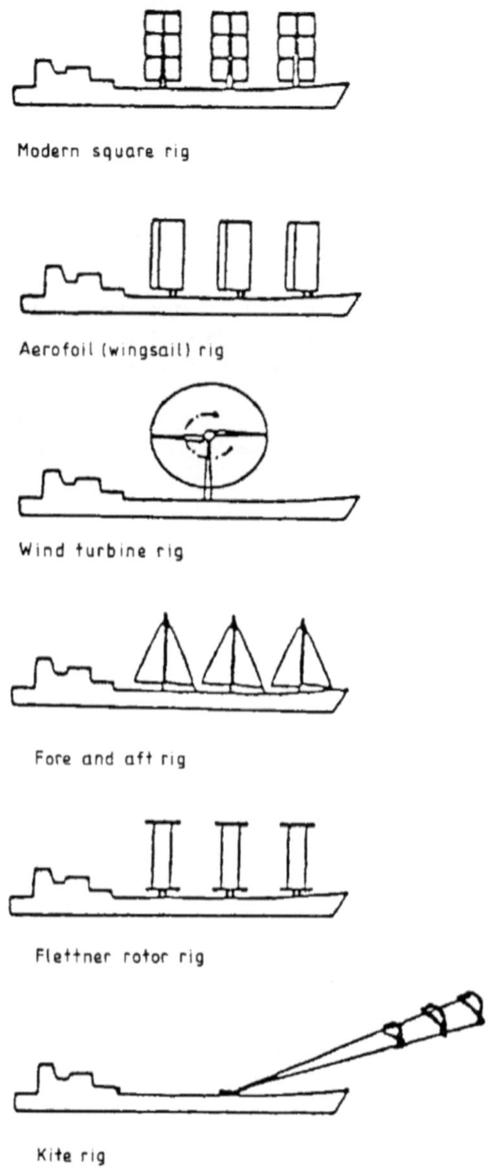

Figure 2: *Main configurations proposed for wind-assisted ships in the 1980s, taken from [15].*

Over the past decade, most configurations have gone from concepts to now being commercially operational, as shown in Figure 3 (based on data edited from [18]). This highlights the ever-growing demand and implementation of wind-assisted commercial vessels. Furthermore, Figure 3 reveals that, for large vessels (greater than 10,000 DWT), sails, rotors and kites are currently the most common configurations. This justifies the particular focus of this paper on these three configurations.

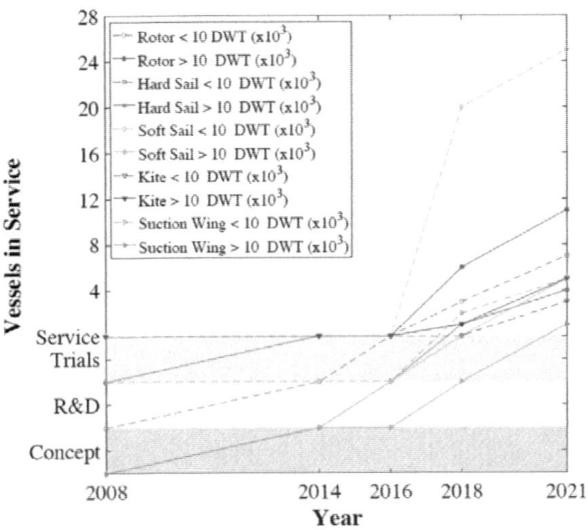

Figure 3: *Development in wind-assisted technologies since 2008.*

3. PRIMARY TECHNOLOGIES

3.1 SAILS

Sailing has historically been central to the development of civilisations and trade, with evidence of sailing vessels dating back as far as the VIth millennia BC [19]. It remained the primary mode of maritime propulsion before being phased out with the development of engine, providing greater power and guaranteed speeds. Since then, sailing has become vastly more complex with yachts being developed to compete in races. This has leads to significant research into the design and performance of sails and sailing yachts, which now proves vital to refine the design and optimize the performance of WASP configurations [20].

Here, the term sails encapsulates soft sails, but also rigid wings, whether using airfoil [21] or the increasingly popular circular arcs section [22, 23, 24, 25]. From a fluid dynamics point of view, a significant distinction needs to be made between sails where the flow remains attached, and those experiencing separated flow. Low camber sails at low angles of attack feature largely attached flow. This can easily be modelled using inviscid codes, and has successfully been implemented since the 1960s [26, 27] and employed in the America's Cup [28]. On the other hand, wings experiencing separated flow regions cannot be analysed with such low order methods. Consequently, either experiments or CFD must be employed. For the latter, it was not until the 1990s that the use of Reynolds-averaged Navier-Stokes CFD could be applied to such sails [29].

Rigid sails and wings have been shown to generate greater lift coefficients than soft sails. In all cases, a greater aspect ratio is associated with better performance, but this leads to a raise in the vertical centre of effort, and thus a larger heeling moment. To reduce the negative effect on stability while maintaining a high sail area and therefore power, multi-masted configurations are considered, with some designs having up to nine masts, as depicted in Figure 4.

Figure 4: *University of Tokyo Wind Challenger, taken from [30].*

As multiple masts is an uncommon configuration on racing yachts, new research investigating the interaction between wings have been conducted specifically for wind-assisted ships. This is for instance the case of Bordogna [24] who conducted experiments with two wings (and rotors), and Macklin [21], who investigated the interaction between two foils numerically. However, there remains research questions when looking at higher number of wings interacting with each other, that are yet unanswered in the literature. Furthermore, current research has mostly focussed on wings behind each other, when new concepts such as that of *Neoline* feature two rows of two masts, equipped with two sails each (a main and a jib).

3.2 ROTATING CYLINDERS

Rotating cylinders comprise two main technologies, namely Flettner rotors and boundary layer suction devices (e.g. *Ventifoil*), with the former being of primary focus in the literature and this paper. Flettner rotors were developed in the 1920s as an alternative to sails. It comprises a vertical spinning rotor which employs the Magnus effect to generate thrust [31]. The concept was first demonstrated in 1926 when a vessel fitted with two Flettner rotors, depicted in Figure 5, crossed the Atlantic [32]. On the other hand, the boundary layer suction devices are a relatively recent development compared to the Flettner rotor. The first use of such sysrem to propel a vessel occurred in the 1980s with *Turbosail*. These devices also utilise the Magnus effect to produce thrust, this time through the use of a suction fan.

Rotors benefit from a higher aspect ratio, albeit with the same stability drawback as sails. However, the tip vortex losses can be alleviated with the use of a top plate, or Thom disc [33]. Rotors also benefit from their ability to be adjusted to match the direction of the wind by simply changing the rotational speed. This helps the device utilise the wind in both legs of a trip which is not always possible for some other wind-assisted propulsion methods [34]. Additionally, rotors are available in numerous dimensions and configurations. The training required is also minimal, consisting of mainly monitoring the system. These reasons make rotors ideal for retrofitting on a range of vessels. It may also explain why, at present, rotors are the most common configuration on vessels above 10,000 DWT, as previously shown in Figure 3.

Figure 5: *Original Flettner rotor ship, taken from [35].*

However, Flettner rotors do suffer from added drag, which can result in an increase in engine power and therefore fuel consumption when the rotors are not operable, for instance close to the wind. This has prompted new developments to mitigate this drag penalty, with the apparition of folding rotor. Nevertheless, not all vessels will have the deck space available for this mechanism to be effective. Another disadvantage is the incompatibility rotors have with container ships due to a lack of deck space. These crafts are particularly important as they are responsible for the majority of emissions due their high speeds [36]. Conversely, rotors are compatible most ship type [37]. New containerized rotor units may also prove a useful solution for container ships, although the overall size and thus power would be limited. Lastly, sails tend to have a higher fuel saving potential (of up to 30%) whereas rotors have an average fuel saving of 8% but a maximum potential of around 20% [7].

3.3. KITES

Although kites are believed to have existed since circa 500 BC, record of their use for propulsion only dates back to the 1820s. The concept of kite powered ships did not actually appear attractive at the time, and was only brought back a few decades ago. A distinction is made between static and dynamic flight for kites [38]. A static flight would be a much more passive system, where the kite acts primarily as a drag generating device, and thus only contributing to the ship's propulsion when sailing close to dead-downwind. Conversely, a dynamic flight would operate in an eight-shape pattern, with a very different kite design allowing lift. As such, a wider range of sailing angles can be achieved, and far greater performance attained [39].

Kites offer significant advantages compared to sails and rotors. Firstly, the ease of installation and ability to be fitted, or retrofitted, to virtually all ships, is substantial. They also offers virtually no reduction in deck space. In addition, because of the atmospheric boundary layer, greater wind speeds are present higher above the water. Kites are able to operate further up from sea level, and thus benefit from these higher wind speeds, which is a squared terms in the lift generation equation. Lastly, the heeling arm generated by a kite is extremely small in comparison to sails and rotors, thereby alleviating the significant stability concerns of the other methods.

However, the technology being more recent and thus less established, coupled with the more restricted wind angles it can operate at, are drawbacks of the kites. The impact on power may also be limited, with research showing a single rotor being more effective than a kite [40]; this will be further evidence in Figure 7 in Section 4.2

4. PERFORMANCE PREDICTION

4.1 HISTORICAL DEVELOPMENTS

The use of velocity prediction programs (VPPs) is another example of performance racing sailing technology cascading down into wind propulsion [20, 41]. Following the pioneering work of the 1930s on yacht performance [42], significant progress was made in the 1970s [43], with numerous developments since the 1990s [44] leading to today's static VPPs [45], but also dynamic ones, accounting for manoeuvres [46, 47].

More recently, VPPs or performance prediction programs (PPPS) as they are more often referred to for wind-assisted ships, have been used to support the optimisation of sails [48], hulls [49] and hydrofoils [50, 51], as well as maximise the performance of both wind-assisted ships [52, 53] and fully wind-powered ships [54]. Similar performance optimisation strategies are employed; for instance, the established use of depowering in yachts [55] has now been applied to wind powered cargo ships [56], albeit with different constraints for the allowable heel angle, much smaller compared to yachts.

4.2 GUIDELINES FOR WASP PERFORMANCE PREDICTION

The performance prediction fundamentally relies on achieving equilibrium for the degrees of freedom (DoFs) considered, out of the 6 DoFs depicted in Figure 6. PPPs for wind-assisted ships typically consider either:

- 3 DoFs (surge, sway, roll);
- 4 DoFs (surge, sway, roll, yaw);
- 6 DoFs.

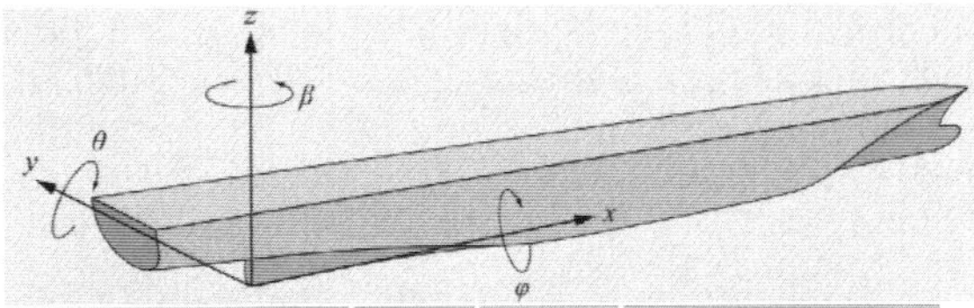

Degree of freedom	Name	Symbol	Sign convention
Linear velocity along the x-axis	Surge	x	Positive forward
Linear velocity along the y-axis	Sway	y	Positive to port
Linear velocity along the z-axis	Heave	z	Positive upwards
Angular motion around the x-axis	Roll	φ	Positive to starboard
Angular motion around the y-axis	Pitch	θ	Positive bow down
Angular motion around the z-axis	Yaw	β	Positive to port

Figure 6: *The 6 degrees of freedom, taken from [37].*

A 4 DoFs PPP is most commonly adopted to yield a reliable performance prediction that allows to accurately ascertain the savings and economic impact of the configuration evaluated. Establishing the constrains on the various degrees of freedom however remains a challenge. The most variation is seen in the critical value of the maximum heel angle, with values as low as 2° [39], recommendations for less than 4° [57], the use of 5° [56], 8° [11] and as high as 10° [58]. Guidelines for constraints applied to the performance prediction of wind-assisted ships are therefore suggested in Table 2.

Table 2: *Guidelines for constraints applied to PPPs for wind-assisted ships.*

Degree of Freedom	Recommended Range
Surge	n/a
Sway	± 5°
Heave	Negligible
Roll	± 4°
Pitch	± 0.5°
Yaw	± 20°

In its simplest form, a 3 DoFs empirical PPP for wind-assisted ships can be developed using the resistance prediction theory of Holtrop and Mennen [59] for the hydrodynamic model. Additionally, the theory inherent to the aerodynamic model is readily available for both yachts [51, 60] and wind-assisted ships [11, 39]. The drive force generated can be subtracted from that provided by the engine, thus leading to a lower engine power and therefore fuel consumption and pollutant emissions for a given speed. The balance of the sail side force created by the wind-assisted system, and the underwater side force, will yield the leeway angle. This is the angle of attack the vessel must adopt in order to generate an equal and opposite side force to that of the wind-assisted configuration. Lastly, the side force applied at the location of the centre of effort results in a heeling moment. To achieve equilibrium, the vessel will roll until a heel angle leading to an equal righting moment is provided.

This would be representative of an early design PPP, allowing for a fast an inexpensive design optimisation. At this stage, it is commonly acknowledged that the performance is not a fully accurate value. Yet, it is appropriate for the purpose of comparative performance prediction [61]. At a later stage of the design, a more advanced PPP may be developed [58], often relying on both hydrodynamic tank testing [57] and aerodynamic wind tunnel testing [24], or CFD [21, 62]. This is crucial to the estimation of fuel savings and comparing the various systems available. The meta-analysis of Neilssen et al. [63] yielded the results presented in Figure 7 for a range of vessel types and cargo sizes, fitted with either a rotor, rigid sail or kite.

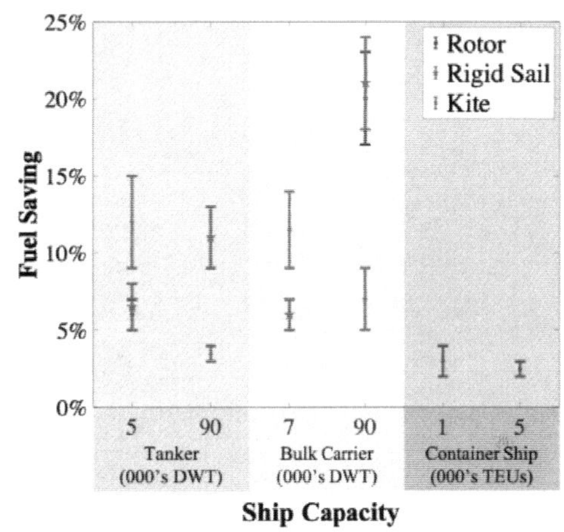

Figure 7: *Potential fuel savings thanks to wind-assisted ship propulsion.*

5. FUTURE STAKES

5.1 PERFORMANCE VALIDATION

The development of aerodynamic coefficients, whether numerically or experimentally remains an area of primary focus. However, the multi-masted configurations, prompted by benefits such as better stability, is driving further work on the interaction between multiples wings and rotors [21, 24, 62], to help support performance prediction.

While such prediction are now well-developed, validation data remain very scarce, not always fully relevant, and ultimately yield a much higher uncertainty compared to that found on yachts [58, 60]. Indeed, as few vessels are currently in service, it will take time for data to be gathered for the array of wind-assisted systems available. Additionally, delays are expected between the acquisition of the data and its availability in the public domain. Because of the industrial interest in the performance prediction, some, if not most of the data, may never be made publically available. It therefore appears critical, and most beneficial from an environmental point of view, to ensure synergies between academia and industry. In the meantime, free-sailing scaled models, such as the *Wallenius Oceanbird*, can provide valuable insights and contribute to the refinement of PPPs [64].

5.2 ENVIRONMENTAL AND ECONOMIC IMPACT

Despite the statistical data on wind strength and direction available as part of the EEDI regulation, weather remains impossible to predict, and as such the benefits of wind-assisted ships remain difficult to precisely ascertain [65]. A comparison of the estimated propulsion savings for a *Maersk Pelican* fitted with a rotor is presented in Figure 8. This shows the output from the PPP of Reche-Villanova et al. [58], the Norsepower (NP) estimation, and sea trial data, highlighting the challenges remaining to predict performance, as well as environmental and economic impact.

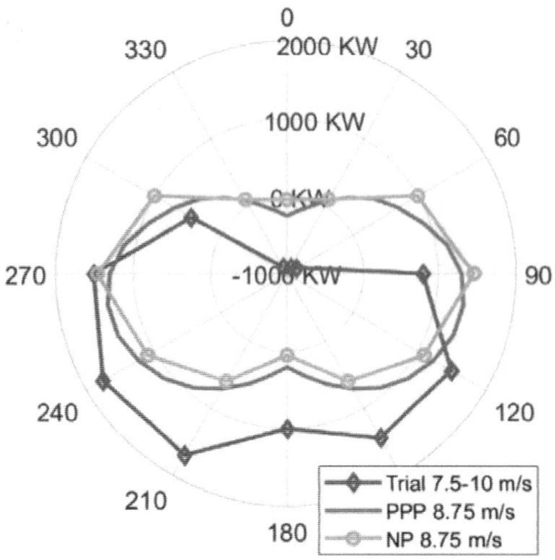

Figure 8: *Comparison between PPP, company (Norsepower) estimate and sea trials, taken from [58].*

With reliable wind data, the power savings estimate can then confidently be converted into reductions in emissions, and the economic impact quantified. This is a vital financial consideration to assess return on investment for wind-assisted ships. Oils prices are also known to have an effect on interest in more sustainable propulsion options, but their volatility is unpredictable. An argument could therefore be made for non-negligible financial incentives to support WASP, making it more financially attractive.

5.3 STRUCTURAL DESIGN AND OPTIMIZATION

The growth of wind-assisted technology has also prompted classification society have developed new rules and regulations intended at wind-assisted ships and their structure. Class NK first released a *Guidelines for Wind-Assisted Propulsion Systems for Ships* in 2019 [66], shortly followed by DNV GL's standard for *Wind-Assisted Propulsion Systems* also in 2019 [67]. In 2020, ABS detailed its requirements for *Wind-Assisted Propulsion System Installation* [68], while the *Rules for Sail Assisted Ships* [69] developed by LR identified basic structural requirements for the masts, posts and supporting structures. More recently, in 2021, BV's *Wind Propulsion Systems* [70] grants additional classification to vessels equipped with wind-assisted propulsion systems, divided between standing parts only (WPS1) and standing and running parts (WPS2). The design combines environmental (wind, sea-state and snow and ice), operating (sailing and out of operation) and system (intact and accidental) conditions. The interface between the ship and rigging is also considered, with a focus on local ship reinforcement as well as global hull girder strength. Indeed, longitudinal strength requirement can be far greater than the conventional wave global loads (peak and trough landing) due to the compressive forces exerted by the rigging. There is, therefore, a need for the vessel-tailored support towers to be fully integrated to the ships structure [71].

With the forecasted growth for wind propulsion, and as greater design experience and sea trial/operation data becomes available, it is expected the scope of the inherent structural regulations will be extended and refined. Moreover, as in many areas, wind-assisted technologies can benefit from the knowledge acquired in sailing yachts structures [72], which features detailed regulations for rig loads and attachments [73, 74], for which the same level of depth remains to be attained on wind-assisted propulsion systems.

6. CONCLUSIONS

Wind-assisted ship propulsion appears as an undeniable part of the future of maritime transportation, offering the greatest long-term potential for emission reduction, as highlighted in this paper. Furthermore, several vessels, utilizing an array of wind-assisted propulsion systems, are already in service, with significant new projects in developments.

The main considerations for sails, rotors and kites have been reviewed. The theory underpinning performance prediction programs was presented, with recommendation on the applicable constraints for all degrees of freedom. The validation of such performance predictions however remains an area of future work, with greater sea trial data available in the public domain being necessary. This constitutes one of the future stakes for wind-assisted ship propulsion, as is the accurate quantification of environmental and economic impact, and the currently under-considered structural implications.

7. ACKNOWLEDGEMENTS

The authors would like to express their thanks to the Institution of Mechanical Engineers (IMechE) for the funding awarded to this research by the Educational Awards Committee.

8. REFERENCES

[1] INTERNATIONAL CHAMBER OF SHIPPING, "Catalysing the fourth propulsive revolution," International Chamber of Shipping, London, 2020.

[2] VAN THEMAAT, W., AND REUDER, B., "Treaty on the functioning of the European Union (Tfeu)," European Competition Law, pp. 25-26, 2018.

[3] BAXTER, J., "Accelerating decarbonisation in shipping: a no regrets approach using wind power," Institution of Mechanical Engineers, Lodnon, 2021.

[4] INTERNATIONAL MARITIME ORGANIZATION, "International Convention for the Prevention of Pollution from Ships (MARPOL)," International Maritime Organization, London, 2020.

[5] INTERNATIONAL TRANSPORT FORUM, "Decarbonising maritime transport: pathways to zer-carbon shipping by 2035," International Transport Forum, Paris, 2018.

[6] SMITH, T., RAUCCI, C., HAJI HOSSEINLOO, S., ROJON, I., CALLEYA, J., SUÁREZ DE LA FUENTE,S., WU, P. AND PALMER, K., "CO_2 emissions from international shipping: Possible reduction targets and their associated pathways," UMAS, London, 2016.

[7] INTERNATIONAL TRANSPORT FORUM, "Navigating towards cleaner maritime shipping," International Transport Forum Policy Papers, Paris, 2020.

[8] ALLWRIGHT, G., "Re-wind, not rewind," Green Shipping 2021, pp. 17-18, January 2021.

[9] ATKINSON, G., NGUYEN, H., AND BINNS, J., "Considerations regarding the use of rigid sails on modern powered ships," Cogent Engineering, vol. 5, no. 1, 2018. https://doi.org/10.1080/23311916.2018.1543564.

[10] BOUMAN, E. A., LINDSTAD, E., RIALLAND, A. I. AND STRØMMAN, A. H., "State-of-the-art technologies, measures, and potential for reducing GHG emissions from shipping – A review," Transportation Research Part D, 52, pp. 408-421, 2017. https://doi.org/10.1016/j.trd.2017.03.022.

[11] TILLIG, F., AND RINGSBERG, J. W., "Design, operation and analysis of wind-assisted cargo ships," Ocean Engineering, 221, pp. 107603, 2020. https://doi.org/10.1016/j.oceaneng.2020.107603.

[12] LINDSTAD H., AND ESKELAND, G. S., "Low carbon maritime transport: How speed, size and slenderness amounts to substantial capital energy substitution," Transport. Res. Part D, 41, pp. 244-256, 2015. https://doi.org/10.1016/j.trd.2015.10.006

[13] TILLIG, F., MAO, W., AND RINGSBERG, J. W., "Systems modelling for energy-efficient shipping," Department of Shipping and Marine Technology, Chalmers University of Technology, Göteborg, 2015. ISSN: 1652-9189.

[14] HOSSAIN, M. J., CHOWDHURY, J. I., BALTA-OZKAN, N., ASFAND, F., SAADON, S., AND IMRAN, M., "Design optimization of supercritical carbon dioxide (s-CO_2) cycles for waste heat recovery from marine engines," Journal of Energy Resources Technology, pp. 1-44, 2021. https://doi.org/10.1115/1.4050006

[15] CLAYTON, B. R., "Wind-Assisted Ship Propulsion," Physics in Technology, 18, pp. 53-60, 1987.

[16] JULIA, E., TILLIG, F., AND RINGSBERG, J. W., "Concept design and performance evaluation of a fossil free operated cargo ship with unlimited range," Sustainability, 1(6), 6609, 2020. https://doi.org/10.3390/su12166609.

[17] THE NAVAL ARCHITECT, "A giant sal for a ship," Green Shipping pp. 22-23, January 2021.

[18] ALLWRIGHT, G., "wind assisted propulsion: challenges and perspectives," KU Leuven Webinar, Leuven, 2020.

[19] CARTER, R., "Boat remains and maritime trade in the Persian Gulf during the sixth and fifth millennia BC," Antiquity, 80(307), pp. 52-63, 2006. https://doi.org/10.1017/S0003598X0009325X.

[20] DET NORSKE VERITAS GERMANISCHER LLOYDS, "Assessment of selected alternative fuelds and technologies," Det Norske Veritas Germanischer Lloyds, 2018.

[21] MACKLIN, J., "Computational fluid dynamics investigation into multiple airfoil interaction to enhance the performance of wind assisted ships," MEng Thesis, Aston Univeristy, Birmingham, 2021.

[22] SOUPPEZ, J.-B. R. G. AND VIOLA, I. M., "Circularc are aerodynamics and applications to downwind yacht sails and wind assisted ships," Marine 2021: IX International Conference on Computational Methods in Marine Engineering, Edinburgh, 2021.

[23] SOUPPEZ, J.-B. R. G., ARREDONDO-GALEANA, A., AND VIOLA, I. M., "Recent advances in numerical and experimental downwind sail aerodynamics," Journal of Sailing Technology, 4(1), pp. 45-65, 2019. https://doi.org/10.5957/jst.2019.4.1.45.

[24] BORDOGNA, G., "Aerodynamics of wind-assisted ships: interaction effects on the aerodynamic

performance of multiple wind-propulsion systems," PhD Thesis, TU Delft, Delft, 2020.

[25] SOUPPEZ, J.-B. R. G., AND VIOLA, I. M., "Aerodynamics of highly cambered circular arcs with a sharp leading edge at low reynolds numbers," Bulletin of the American Physical Society, 73rd Annual Meeting of the APS Division of Fluid Dynamics, Chicago, 2020.

[26] MILGRAM, J. H., "The Aerodynamics of sails," in 7th Syposium on Naval Hydrodynamics, 1968.

[27] GENTRY, A., "The aerodynamics of sail interaction," 3rd AIAA Symposium on Sailing, California, 1971.

[28] GENTRY, A., "The application of computational fluid dynamics to sails," Proceedings of the Symposium of Hydrofynamic Performance Enhancement for Marine Applications, 1988.

[29] HEDGES, K., RICHARDS, P., AND MALLISON, G., "Computer modelling of downwind sails," Journal of Wind Engineering and Industrial Aerodynamics, 63, pp. 95-110, 1996.

[30] OUCHI, K., UZAWA, K., KANAI, A. AND KATORI, M., "Wind challenger: the next generation hybrid sailing vessel," 3rd International Symposium on Marine Propulsors, 2013.

[31] SEIFERT, J., "A review of the Magnus effect in aeronautics," Progress in Aerospace Sciences, 55, pp. 17-45, 2012. https://doi.org/10.1016/j.paerosci.2012.07.001

[32] NUTTALL, P., AND KAITU'U, J., "The Magnus effect and the Flettner rotor: potential application for future oceanic shipping," Journal of Pacific Studies, 36(2), pp 162-183, 2016.

[33] THOM, A., "Effect of discs on the air forces on a rotating cylinder," Aerodynamic Research Council R&M 1623, 1934.

[34] SEARCY, T. "Harnessing the wind: A case study of applying Flettner rotor technology to achieve fuel and cost savings for Fiji's domestic shipping industry," Marine Policy, 86, pp. 164-172, 2017. https://doi.org/10.1016/j.marpol.2017.09.020

[35] CRAFT, T. J., IACOVIDES, H., JOHNSON, N., AND LAUNDER, B. E., "Back to the future: Flettner-Thom rotors for maritime propulsion?," Tubulence, Heat and Mass Transfer, 2012.

[36] BURDEN, A., LLOYD, T., MOCKLER, S., MORTOLA, L., SHIN, I. B., AND SMITH, B., "Concept design of a fast sail assisted feeder container ship," University of Southampton Report, Southampton, 2010.

[37] SOUPPEZ, J.-B. R. G., "Ships and Maritime Transportation," in Grote, K.-H. and Hefazi, H. (Eds.), Springer Handbook of Mechanical Engineering, Springer Handbooks, pp. 1139-1164, 2021. 10.1007/978-3-030-47035-7_25.

[38] CADELEN, B., GRIFFON, F., LANUSSE, P., SABATIER, J. AND PARLIER, Y., "Modelling and control of a tethered kite in dynamic flight," Journal of Sailing Technology, 3(1), pp. 1-24, 2018. https://doi.org/10.5957.jst.2018.03.

[39] PECK, B., "Development of a velocity prediction program to investigate and integrate wind assisted ship propulsion technologies with a focus on kites," MEng Thesis, Aston University, Birmingham, 2021.

[40] TRAUT, M., GILBERT, P., WALSH, C., BOWS, A. FILIPPONE, A., STANSBY, P., AND WOOD, R., "Propulsive power contribution of a kite and a Flettner rotor on selected shipping routes," Applied Energy, 113, pp. 362-372, 2014. https://doi.org/10.1016/j.apenergy.2013.07.026

[41] EGGERS, R., "Prediction methods for team akzonobel in the volvo ocean race and wind assisted ship propulsion for merchant vessels," 25th International HISWA Symposium on Yacht Design and Yacht Construction, Amsterdam, 2018.

[42] DAVIDSON, K. S. M., "Some experimental studies of the sailing yacht," Society of Naval Architects and Marine Engineers, 1936.

[43] KERWIN, J. E., "A velocity prediction program for ocean racing yachts," SNAME report 75-17, 1978.

[44] VAN OOSSANEN, P., "Predicting the speed of sailing yachts," SNAME transactions 101, 1993.

[45] OFFSHORE RACING CONGRESS, "ORC VPP Documentation," Offshore Racing Congress, 2019.

[46] RIDDER, E. J., VERMEULEN, K. J., AND KEUNING, J. A., "A Mathematical model for the tacking maneuver of a sailing yacht," 18th International HISWA Symposium on Yacht, Amsterdam, 2004.

[47] BATTISTIN, D. AND LEDRI, M., "A tool for time dependent performance prediction and optimization of sailing yachts," 18th Chesapeake Sailing Yacht Symposium, Annapolis, Maryland, 2007 .

[48] COURTEL, P., "Preliminary sail shape optimisation using neural network aerodynamic and hydrodynamic models," EMship Master Thesis, Rostock, 2020.

[49] SOUPPEZ, J.-B. R. G., "On the applications of modern naval architecture techniques to historical crafts," Historic Ships, 1-15, 2016.

[50] SOUPPEZ, J.-B. R. G., DEWAVRIN, J. M. M.-A., GOHIER, F. AND BORBA LABI, G., "Hydrofoil configuration for sailing superyachts: hydrodynamics, stability and performance," in Design & Construction of Super & Mega Yachts, Genoa, 2019.

[51] BORBA LABI, G., "Velocity prediction program development for hydrofoil-assisted sailing monohulls," MSc Thesis, University of Liege, Liege, 2019.

[52] VIOLA, I. M, SACHER, M., XU, J. AND WANG, F., "A numerical method for the design of ships with wind-assisted propulsion," Ocean Engineering, 105, pp. 33-42, 2015. https://doi.org/10.1016/j.oceaneng.2015.06.009.

[53] MALMEK, K., DHOME, U., LARSSON, L., WERNER, S., RINGSBERG, J. W., AND FINNSGARD, C., "Comparison of two rapid numerical methods for predicting the performance of multiple rigid wind-sails," 5th International Conference on Innovation in High Performance Sailing Yachts and Sail-Assisted Ship Propulsion, Gothenburg, 2020.

[54] LINDEN, V., AND SOUPPEZ, J.-B. R. G., "Sailing towards sustainable trading with wooden cargo schooner," in British Conference of Undergraduate Research, Sheffield, 2018.

[55] HANSE, H., RICHARDS P. J., AND JACKSON, P., "Reef and flat or twist and ease: An improved trim parameter model for yacht sails," International Journal of Small Craft Technology, 2010.

[56] OLSSON, F., MARIMON GIOVANNETTI, L., WERNER, S., AND FINNSGARD, C., "A performance depowering investigation for wind powered cargo ships along a route," Journal of Sailing Technology, 5(1), pp. 47-60, 2020. https://doi.org/10.5957/jst/2020.3.1.47.

[57] VAN DER KOLK, N., BORDOGNA, G., MASON, J. C., DESPRAIRIES, P., AND VRIJDAG, A., "Case study: wind-assisted ship propulsion performance prediction, routing and economic modelling," International Conference on Power and Propulsion Alternatives for Ships, The Royal Insitution of Naval Architects, London, 2019.

[58] RECHE-VILANOVA, M., HANSEN H., AND BINGHAM, H. B., "Performance prediction program for wind-assisted cargo ships," Journal of Sailing Technology, 6(1), pp. 91-117, 2021. https://doi.org/10.5957/jst/2021.6.1.91.

[59] HOLTROP, J., AND MENNEN, G., "An approximate power prediction method," International Shipbuilding Progress, vol. 29, no. 335, pp. 166-170, 1982.

[60] SOUPPEZ, J.-B. R. G., "Development and validation of a computational fluid dynamics hydrodynamic model of the Stewart 34 for velocity prediction program applications," MEng Thesis, The University of Auckland, Auckland, 2014. 10.13140/RG.2.2.12456.26884/1.

[61] THOMAS, J. E., AND SOUPPEZ, J.-B. R. G., "Comparative Performance Prediction of Historical Thames A Rater Class Designs," Historic Ships, pp. 67-76, 2018.

[62] KHAN, L., "Wind assisted propulsion alternatives for commercial shipping - a computational fluid dynamics study of Flettner rotors," MEng Thesis, Aston University, Birmingham, 2021.

[63] NEILSSEN, D., TRAUT, M., KOHLER, J., MAO, W., FABER, J. AND AHDOUR, S., "Study on the analysis of market potentials and market barriers for wind propulsion technologies for ships," CE Delft, Delft, 2016.

[64] KVARNEFALK, A. AND ROBSARVE, O., "Mätning av rigglaster påsegelbåten Christiane," KTH Royal Institute of Technology, 2021.

[65] CHOU, T., KOSMAS, V., ACCIARO, M. AND RENKEN, K., "A comeback of wind power in shipping: an economic and operational review on the wind-assisted ship propulsion technology," Sustainability, 13(4), 2021. https://doi.org/10.3390/su13041880

[66] CLASSNK, "Guidelines for wind-assisted propulsion systems for ships," ClassNK, 2019.

[67] DET NORSKE VERITAS GERMANISCHER LLOYDS, "Wind assisted propulsion systems," Det Norske Veritas Germanischer Lloyds, 2019.

[68] AMERICAN BUREAU OF SHIPPING, "Wind assisted propulsion systems instalation," American Bureau of Shipping, Spring, Texas, 2020.

[69] LLOYDS REGISTER, "Rules for sail assisted ships," Lloyds Register, London, 2020.

[70] BUREAU VERITAS, "Wind propulsion systems, Rule note NR206 DT R01 E," Bureau Veritas, Paris, 2021.

[71] PAAKKARI, V., "Modern rotor sail technolohy helps ships save fuel and reduce emissions - performance and experiences from recent installations," RINA Power & Propulsion Alternatives for Ships, London, 2019.

[72] SOUPPEZ, J.-B. R. G., "Structural design of high performance composite sailing yachts under the new BS EN ISO 12215-5," Journal of Sailing Technology, 3(1), pp. 1-18, 2018. https://doi.org/10.5957.jst.2018.02.

[73] DET NORSKE VERITAS GERMANISCHER LLOYDS, "Design and construction of large modern yacht rigs," Det Norske Veritas Germanischer Lloyds., Baerum, 2016.

[74] INTERNATIONAL ORGANIZATION FOR STANDARDIZATION, "ISO 12215-10:2020 Small craft - Hull construction and scantlings - Part 10: Rig loads and rig attachment in sailing craft," International Organization for Standardization, Geneva, 2020.

9. AUTHORS BIOGRAPHY

Lukman Khan is a graduate of the MEng Mechanical Engineering at Aston University, and the author of a thesis entitled: *Wind Assisted Propulsion Alternatives for Commercial Shipping - A Computational fluid Dynamics Study of Flettner Rotors'.*

Joseph J. R. Macklin is a graduate of the MEng Mechanical Engineering at Aston University, and the author of a thesis entitled: *'Computational Fluid Dynamics Investigation into Multiple Airfoil Interaction to Enhance the Performance of Wind Assisted Ships'*.

Benjamin C. D. Peck is a graduate of the MEng Mechanical Engineering at Aston University, and the author of a thesis entitled: *'Development of a Velocity Prediction Program to Investigate and Integrate Wind Assisted Ship Propulsion Technologies with a Focus on Kites'*.

Owen Morton is a graduate of the BEng Mechanical Engineering at Aston University, and the author of dissertation entitled: *'The development of a velocity prediction program to compare different methods of wind assisted cargo ships'*.

Jean-Baptiste R. G. Souppez is the Senior Teaching Fellow in Mechanical Engineering and Design at Aston University. He is also the UK Principal Expert in Small Craft Structures, and the Deputy Editor-in-Chief of the Society of Naval Architects & Marine Engineers (SNAME) Journal of Sailing Technology. His research in fluid dynamics features a wide range of experimental methods supporting his award-winning work on the aerodynamics of downwind racing yachts sails.

PERFORMANCE PREDICTION AND DESIGN OF WIND-ASSISTED PROPULSION SYSTEMS

F Thies and **J W Ringsberg**, Chalmers University of Technology, Sweden

SUMMARY

Wind-assisted propulsion is seen as one of the main alternatives to potentially achieve large emission reductions in shipping. However, wind-assisted propulsion introduces new challenges in the design, retrofitting and performance prediction as well as the performance analysis. This paper presents and compares methods to predict the performance of wind-assisted propulsion, using the validated performance prediction model ShipCLEAN. Focus is put on evaluating the difference between 1 degree of freedom (1 DOF) and 4 DOF methods as well as the impact of aerodynamic interaction effects in between multiple sails. Practical design considerations and performance differences are discussed based on an example ship. The study includes a comparison of the performance of different sail types (Flettner rotors, Wing sails and Suction wings) under realistic operational conditions.

NOMENCLATURE

[Symbol]	[Definition] [(Unit)]
B	Beam (m)
DOF	Degree of freedom (-)
FP	Forward perpendicular (-)
L_{oa}	Length over all (m)
L_{pp}	Length between perpendiculars (m)
T	Draft (m)
TWA	True wind angle (deg)
TWS	True wind speed (kn, m/s)
$WASP$	Wind-assisted propulsion (-)
v_S	Ship speed (kn)
Δ	Displacement (t)

1. INTRODUCTION

Shipping today accounts for 90% of all freight transport. Forecasts of the world's transportation needs in 2050 show that it will double the current level. At the same time, shipping must become more energy-efficient and before 2050 reduce its contribution to emission of greenhouse gases to 50% of the level in 2008 (IMO, 2018). A first, and crucial step towards reducing the fuel consumption of ships is to be able to accurately predict the consumption of existing ships as well as the positive impact on the reduction of fuel consumption due to: (i) operational or retrofitting measures, (ii) design changes, and (iii) the installation of energy-saving devices such as sails. To achieve this, each part, and the interaction of all parts in a ship's energy system must be understood.

The ShipCLEAN simulation model presented by Tillig (2020) is developed as a coupled model including ship performance prediction and maritime transport logistics. Its capability and accuracy in ship performance predictions have been validated for commercial ships with and without Flettner rotors. It is used in the current study to compare methods to predict the performance of three sails, i.e., wind-assisted propulsion technologies, under realistic operational conditions: Flettner rotors, Wing sails, and Suction wings.

2. PERFORMANCE PREDICTION

With the increasing interest in wind-assisted propulsion, the need for accurate and versatile prediction models increases simultaneously. One basic question is the level of accuracy that is needed for a prediction of the fuel savings achieved by wind-assisted propulsion. In this section some important effects, hydrodynamically and aerodynamically, and their importance for an accurate performance prediction will be discussed.

2.1 DEGREES OF FREEDOM

For traditionally propelled ships, the typical performance prediction methods only include the resistance (including resistance from waves and wind) and the propulsive efficiency; hence they are 1 degree of freedom (1 DOF). Such an approach was also applied to wind-assisted ships, i.e., the sails would be treated as an additional propulsor creating pure thrust. However, since the sails create large side forces, more degrees of freedom must be considered, both to accurately predict the added resistance created from drifting and steering but also to model constraints for the sail forces, i.e., maximum rudder or heel angles. Thus, it is crucial to include at least 4 DOF: surge (thrust), drift (side force), yaw (side force), and heel (side force). The dynamical sinkage of the ship is only dependent on the ship's speed and will thus not change by the addition of wind propulsion unless the systems introduce a vertical force which is not the case for the wind propulsion systems discussed here.

As an example, one can estimate the necessary propeller thrust for a ferry, which is presented in more detail in Section 4.2. With a TWA of 60 deg, TWS of 20 kn, wave height of 1 m and a ship speed of 16 kn, the ship has an estimated total resistance (including added resistances) of 463.6 kN. The sail thrust of one 5×30 m Flettner rotor in this condition is estimated to be 55.3 kN, i.e., about 12% of the total required thrust. However, considering 4 DOF, a drift and rudder resistance of about 8.3 kN must be added, reducing the thrust reduction to about 10%. Additionally, the sail forces would be slightly reduced since the ship drifts with about 1.5 deg. The example shows the importance of considering 4 DOF, even for WASP ships with small sail areas. It gets even more important for ships with larger sail areas and for conditions closer to the wind, when sails often must be reefed for the ship to be able to hold course and maneuver (Tillig, 2020).

2.2 AERODYNAMIC INTERACTION

Sails on deck of a ship are interacting with each other and the deck or superstructure. Recently, numerous studies focused on the interaction effects of sails on a ship, e.g., Bordogna et al. (2020), Tillig and Ringsberg (2020). The results of these studies show that, in-between sails, potential flow interaction effects are predominant. Those effects are caused by the bound and tip/root vortices of the sails and cause a variation of local wind speed and direction in the flow field, i.e., each sail will experience a wind speed and direction different from the free flow speed and direction and different from the other sails. These induced speeds in a field of multiple sails cause a variation of the force centre of the combined force of all ships but no large losses. In general, the combined force acts more forward in a field of sails with modelled interaction than in the same field when interactions are disregarded. Thus, aerodynamical interactions are crucial to model to accurately predict the jaw moments and thus the necessary rudder angle and for reliable sail control, as discussed in the following section. The thrust forces from individual sails with and without interactions are compared in Figure 1.

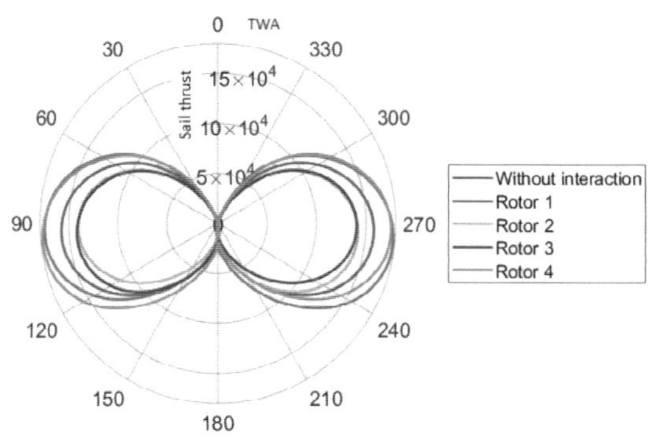

Figure 1: Effects of sail interaction (Tillig and Ringsberg, 2020).

2.3 SAIL CONTROL

In 1 DOF simulation model predictions, sails are often regarded to act in the trim delivering the maximum thrust. However, when 4 DOF are modelled, the sails shall be trimmed to find the best balance between thrust, drift, heel, and rudder angle as well as the associated added resistances. Further, constraints on the rudder and heel angle shall be kept and thus the sail might have to be de-powered (reefed).

Considering the complexity and the coupling between ship dynamics, sails, interaction effects, and the dependency of the rudder forces on the propeller thrust (and thus sail thrust/trim), sail trim becomes a classic optimization problem that is not analytically solvable. Further, especially in downwind conditions, it can be beneficial to inverse the aft sails (e.g., reverse the rotation of Flettner rotors), creating less thrust or even slightly negative thrust but acting as a rudder to counter the jaw forces of the more efficient forward sails. These effects only occur on ships with large sail areas or fully sail-powered ships. For more modest wind-assisted ships, the main benefit of sail trim optimization is found in the head wind range, as illustrated in Figure 2.

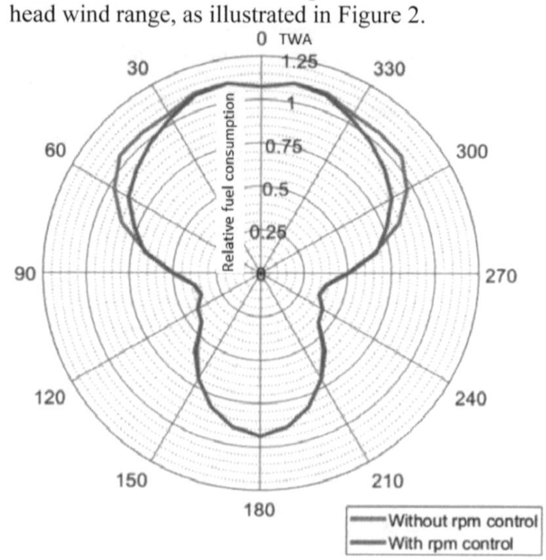

Figure 2: Fuel savings of 4 Flettner rotors on a RoRo ship, with and without optimized sail trim (Tillig and Ringsberg, 2020).

3. SHIPCLEAN

ShipCLEAN is a generic ship energy systems model, i.e., a ship performance prediction model that is modularized by means of the energy systems on board a ship. The structure and details of the model are extensively published and summarized in Tillig (2020). ShipCLEAN consists of two main parts. Firstly, a 1 DOF, generic power prediction model which provides power predictions for any cargo ship while requiring only the main parameters as input. Secondly, a 4 DOF part which includes environmental influences as well as methods for sails and aerodynamic interaction. The expected and achieved accuracy in predictions using this model is presented in Tillig et al. (2018) and Tillig (2020). In this study, both parts of ShipCLEAN are used.

4. DESIGN EXAMPLE

The example ship is a ferry with main dimensions according to Table 1. The ferry is equipped with one 5×30m Flettner rotor which is mounted midships, i.e., at $0.5 \times L_{pp}$ from the aft (AP), as illustrated in Figure 3 and three propellers, one center and two steerable Pods. However, for the studies in this work, the ship is assumed to have two conventional high lift rudders, since Pods would not create sufficient yaw moments when the main propulsion thrust is delivered from the sails.

L_{oa}	156.45 m
B	24.8 m
T	5.5 m
Δ	11996 t
Service speed, v_S	16 kn

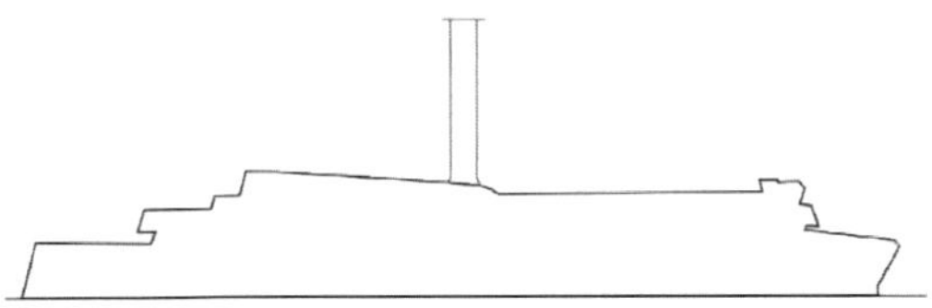

Figure 3: Sketch of the example ferry with one Flettner rotor.

The ferry operates on a route between Rostock (Germany) and Gedser (Denmark) with five daily departures in each direction. The open sea leg of the route is about 23 nm long with a course over ground of 165 deg (bound to Germany) and 345 deg (bound to Denmark). To accurately predict the long-term savings, a statistical approach using historical weather on the route is used, which is described in detail in Tillig and Ringsberg (2020). In short, this approach uses probability distribution functions of the wind speed and direction to estimate weights for the points evaluated in a polar diagram. In that way, the computational time can drastically be reduced while still maintaining a high level of accuracy of the prediction of fuel savings. Long term wind statistics (based on the years 2016 to 2018) on the route are summarized in a wind scatter plot in Figure 4. It is shown that there are two dominating true wind directions, between 190 and 210 deg (southwest) and between 10 and 30 deg (northeast).

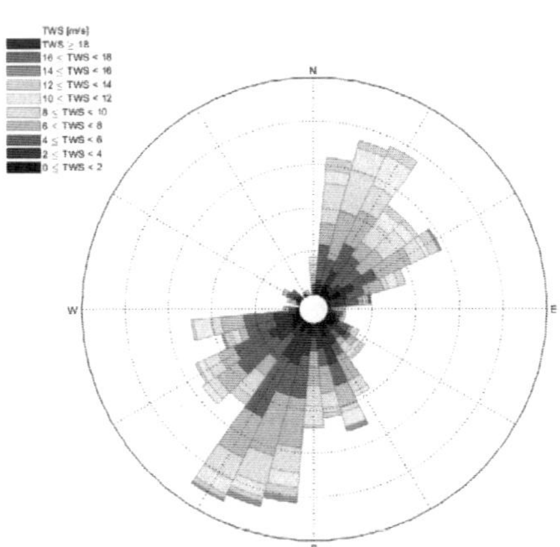

Figure 4: Wind scatter plot based on statistics from 2016 to 2018.

4.1 SAIL POSITION

The longitudinal position of the sail is crucial for the induced yaw moment. The further aft the sail is positioned, the higher the load on the rudder will be. A high rudder load is favourable since the lift to drag ratio of the rudder is much better than the ratio of the hull. However, large rudder angles must be avoided to keep manoeuvrability. Thus, a too far aft positioned sail might be de-powered most of the time. To evaluate the impact of the longitudinal position, the sail is moved from the forward perpendicular to the aft perpendicular in steps of $0.2 \times L_{pp}$. The fuel savings are estimated using the statistical approach discussed above. Results are presented in Figure 5, which shows that a position slightly aft of the midship (50% L_{pp}) is most favourable.

4.2 SAIL AREA AND TYPE OF SAIL

To determine the best sail type and sail area for a ship, three main parameters must be considered: (i) the available space on deck, (ii) height restrictions and (iii) the return of investment, i.e., the achieved savings in comparison to the installation and running costs. In this study, different sail areas of two sail types, Flettner rotors and Suction wings, are compared by means of the

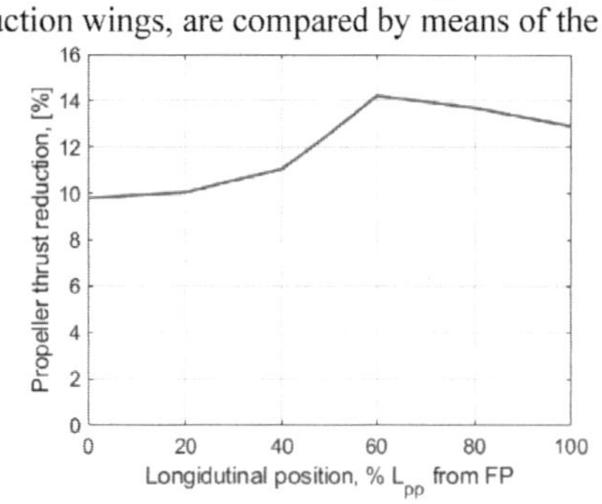

Figure 5: Influence of the longitudinal position on the power savings.

achieved savings. It is assumed that up to six Flettner rotors (5×30 m) and up to six Suction wings (5×30 m and 10×60 m) can be fitted on the ferry. Since installation and running costs are not available to the authors, only the achievable savings will be presented and discussed.

A first comparison of the Flettner rotor and the Suction wing can be done by means of the maximum lift coefficient. While Flettner rotors can achieve lift coefficients of up to 12, the maximum lift coefficient of the Suction wing is proven to be 5. On the other hand, the Suction wing has a higher lift to drag ratio of about 3.5 at the maximum lift point than the Flettner rotor, which has a lift to drag coefficient of about 3 at the highest lift and even at the typical working points. The difference in achievable lift coefficient is the reason for the choice of installed sail areas, where the maximum sail area for the Flettner rotors is predicted to be 6×150 m^2, i.e., 1000 m^2, while the maximum sail area for the Suction sails is 6×300 m^2, i.e., 1800 m^2.

Figure 6 compares propeller thrust reduction if the ship had been equipped with a different number of Flettner rotors or Suction wings. It is shown that the increase of the savings is not linear, especially for the Flettner rotor. While savings are almost similar for one 150 m^2 Flettner rotor and one 300 m^2 Suction wing, 1000 m^2 Flettner rotor (six units) give a saving of about 54%, while 1800 m^2 Suction wing give about 66%. This shows the effect of higher lift to drag ratios of the sail for high sail areas, which is rather unimportant for small sail areas.

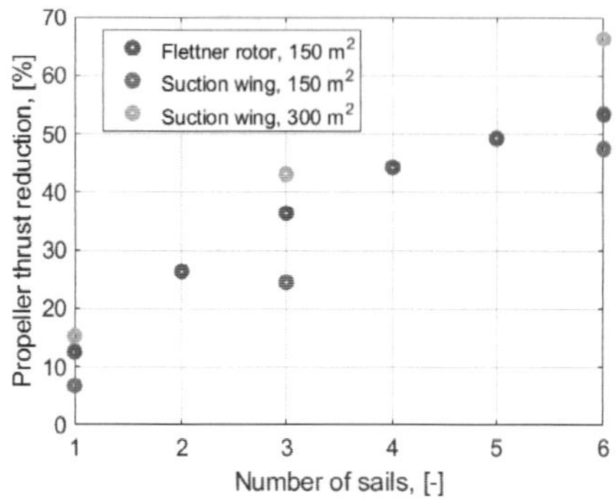

Figure 6: Achievable propeller thrust savings with different number of installed Flettner rotors and Suction wings.

5. CONCLUSIONS

Using a ferry ship as an example, this study discussed the necessity of considering 4 DOF and aerodynamic interaction for an accurate performance prediction for wind-assisted ships. By neglecting drift forces and yaw moments, the total resistance of the ship is under-predicted since added resistances from drift and the rudder are not included. Further, 1 DOF models cannot respect constraints on the rudder and heel angles. A 4 DOF model is also crucial to accurately model the optimal sail trim, considering the constraints and the optimal overall performance, which might involve de-powering the sails to reduce rudder and drift resistance.

In the example with a ferry operating on a short route in the Baltic Sea, it was shown how a statistical approach can be employed to accurately predict the long-term performance of WASP systems. The design of the WASP systems was discussed by varying the position and the size as well as type of the sail. Sails that were positioned optimally showed to achieve about 14% power savings, while those at the least optimal position only achieved 10% power savings. The sail size variation showed that the savings do not increase linearly with the sail size, mainly due to the rudder constraints and the increasing drift resistances.

In the comparison of Flettner rotors and Suction sails, it was shown that similar savings can be achieved with different sail types, given that the sail area is changes accordingly. However, the larger the sail area, the more important the lift to drag ratio. Thus, it must be assumed that the choice of sail type will be more important for fully wind-propelled ships. It was also presented that the savings do not increase linearly, mainly due to drift and rudder angles and maximum rudder angle constraints. Since this effect is more pronounced for the Flettner rotors, it can be concluded that the higher lift to drag ratio of the Suction wing is favourable at large sail areas while the large lift coefficients of the Flettner rotors are favourable for smaller sail areas. The results showcase the difficulty of comparing different sail types. Since the performance of the sails is always dependent on the sail area, it must be concluded that the most suitable parameter for comparison of different sail types would be the payback time. However, such comparison requires the availability of installation and running costs of the sail types.

6. ACKNOWLEDGEMENTS

This study received funding from Chalmers University of Technology Foundation for the strategic research project "Hydro- and aerodynamics". It also received funding from the EU Interreg North Sea Europe program from the project "WASP – Wind Assisted Ship Propulsion; Run wind propulsion technology real life trials on sea going ships in operation, showcase proven concepts, market adaptation, green sea transport" (project no 38-2-6-19; https://northsearegion.eu/wasp/).

7. REFERENCES

[1] Bordogna, G., Muggiasc, S., Giappino, S., Belloli, M., Keuning, J.A., Huijsmans, R.H.M., The effects of the aerodynamic interaction on the performance of two Flettner rotors, Journal of Wind Engineering and Industrial Aerodynamics, 196, DOI: 10.1016/j.jweia.2019.104024, 2020

[2] IMO, GHG-Emissions (Online). Available at: http://www.imo.org/en/OurWork/Environment/PollutionPrevention/AirPollution/Pages/GHG-Emissions.aspx (accessed: 21 08 2021).

[3] Tillig, F., Simulation model of a ship's energy performance and transportation costs. Doctoral thesis (New series, no. 4750). Chalmers University of Technology, Gothenburg, Sweden, 2020

[4] Tillig, F., Ringsberg, J.W., Design, operation and analysis of wind-assisted cargo ships. Ocean Engineering. 211(1): 1-23. (https://doi.org/10.1016/j.oceaneng.2020.107603.), 2020

[5] Tillig F, Ringsberg JW, Mao W, Ramne B., Analysis of uncertainties in the prediction of ships' fuel consumption - from early design to operation conditions. Ships and Offshore Structures. 13(S1): S13-S24. (https://doi.org/10.1080/17445302.2018.1425519.), 2018

8. AUTHORS BIOGRAPHY

Fabian Thies holds the current position of Postdoctoral researcher at Chalmers University of Technology, Department of Mechanics and Maritime Sciences. During his PhD he developed the generic performance model ShipCLEAN. Current and previous work focus on the design and performance prediction of wind-assisted and fully wind-powered ships as well as zero-emission shipping.

Jonas W Ringsberg holds the current position of Professor of Marine Structures at Chalmers University of Technology, Department of Mechanics and Maritime Sciences. He is the Head of the Division of Marine Technology at the Department of Mechanics and Maritime Sciences. His previous experience includes, among others, research on numerical modelling, simulation and analysis of ship energy efficiency, wind-assisted ship propulsion, and fossil-free ship propulsion alternatives.

A CFD STUDY ON WIND ASSISTED PROPULSION TECHNOLOGY FOR COMMERCIAL SHIPPING

W C P Hopes, **D R Pearson** and **J E Buckingham**, BMT Defence and Security UK Ltd, UK

SUMMARY

Wind assisted propulsion has the potential to significantly reduce emissions from global shipping, with a range of technologies available including wing sails, kites, Flettner Rotors, and Suction Aerofoils such as Turbosails. Despite this, there are only a few studies publicly available on the performance of Suction Aerofoils, which operate on the same principle as a traditional sail but use boundary layer suction to develop high lift coefficients. Therefore, a CFD study was carried out in Numeca FINE/Marine software to determine values for the lift and drag coefficients over a range of angles of attack and suction levels. Supplementary studies also investigated the effects of introducing an endplate to reduce end vortices, and the effect of a ship side on the wind speed incident on the Suction Aerofoil. The plots of lift and drag coefficients developed in this paper were used to inform the BMT-led VTAS project [1], as well as contributing to wider knowledge on wind assisted propulsion for shipping. Throughout the study, Suction Aerofoils continued to show promise as an effective means of auxiliary propulsion for ships, and their use should be encouraged as part of the range of solutions to tackle the pressing issue of climate change.

NOMENCLATURE

AoA	Angle of attack (°)
c	Chord of suction aerofoil (m)
C_D	Coefficient of drag (-)
C_L	Coefficient of lift (-)
Cq	Coefficient of suction (-)
CFD	Computational Fluid Dynamics
dx	Width of suction strip (m)
dwt	Deadweight tonnage
EEDI	Energy Efficiency Design Index
EEXI	Energy Efficiency Existing Ship Index
ESTs	Energy saving technologies
GHG	Greenhouse gas
IMO	International Maritime Organization
n	Number of suction strips (-)
ORC	Organic Rankine Cycle
Q	Volumetric flow rate (m^3/s)
Re	Reynold's number (-)
SST	Shear stress transport
U_{suc}	Suction velocity (m/s)
U_∞	Freestream velocity (m/s)
y^+	Non-dimensional wall distance (-)
Z	Height from base of suction aerofoil (m)

1. INTRODUCTION

The need to reduce global CO_2 emissions is urgent and will require widespread innovative approaches to limit the increasing effects of climate change. Despite shipping already being considered one of the most efficient forms of transport, the latest calculation by the International Maritime Organization (IMO) estimated total shipping emissions were 938 million tonnes of CO_2 during 2012 – 2.6% of global emissions for that year [2]. Without further action, it is estimated these emissions could grow between 50% and 250% by 2050 on current trajectories, simply due to predicted growth in world trade [2].

In response to climate change, the IMO's Initial Strategy on Reduction of GHG Emissions from Ships sets out ambitions to reduce the carbon intensity (CO_2 emissions per transport work) of international shipping by at least 40% by 2030, pursuing efforts towards 70% by 2050; and to reduce the total GHG emissions from international shipping by at least 50% by 2050 compared to 2008 [3]. In June 2021, as part of short-term measures to achieve these emission reduction goals, the IMO adopted requirements for an EEXI score to be calculated for existing ships, alongside the EEDI score for new ship builds [4]. These scores are technical measures of how efficient a ship is, and the base calculation is summarised below [5]:

$$EEXI\ or\ EEDI = \frac{impact\ on\ environment}{benefit\ to\ society} = \frac{CO_2\ emissions}{dwt \times speed}$$

With EEXI set to enter into force on 1st January 2023, and subsequent phases requiring tougher EEXI/EEDI scores to be met, many commercial ships will require improvements in fuel efficiency. Therefore, several approaches may be taken to meet the 2023 regulations and ultimately reduce CO_2 emissions – examples include (but are not limited to) the introduction of the following energy saving technologies (ESTs):

- Improving the hydrodynamic performance of the hull, through techniques such as air lubrication and improved propeller performance
- Waste heat recovery e.g. using Organic Rankine Cycles (ORCs)
- Wind assisted propulsion with kites, soft sails, wing sails, Flettner Rotors or Suction Aerofoils

The BMT-led VTAS project [1] concluded that a combination of ESTs such as these have the potential to offer fuel savings of up to 20% if retrofitted to a typical 61,000 dwt bulker [6].

Wind assisted propulsion technologies are particularly suitable to be fitted on new or existing ships, and with high technological maturity they are available today to make significant fuel savings in the short-term whilst alternative fuels and associated infrastructure continue to be developed.

In the medium/long-term, wind assisted propulsion technologies have the key benefit of being compatible with all potential future fuels. Given the uncertainty surrounding which alternative fuel(s) will eventually be adopted by the global shipping fleet, it is anticipated initial uptake for any fuel will be expensive with varying availability between countries. Therefore, reducing the fuel consumption of new fuels through installation of wind assisted propulsion has the potential to make this transition to sustainable fuels more feasible [7].

Whilst wind assisted propulsion technologies range from wing sails, to kites, Flettner Rotors, and Suction Aerofoils, all should be considered complementary and part of the solution to tackling maritime emissions. This paper however focuses on the study of Suction Aerofoils with the following aims:

1. Determine lift and drag coefficients (C_L & C_D) for a Suction Aerofoil over a range of angles of attack and suction levels to inform the BMT-led VTAS project.
2. Contribute to the published knowledge on wind assisted propulsion for shipping to encourage uptake within the industry.

Suction Aerofoils, such as the original Turbosail developed by Jacques Cousteau in the 1980s [8], or the Ventifoil currently being developed by Econowind [9] offer many advantages as wind propulsion technology; much higher lift coefficients are possible (when compared to a traditional 'soft' sail) due to the use of suction on the leeward side, they can be easily automated to minimise any additional training required by a ship's crew, and can be reefed (depowered) in strong winds by reducing the level of suction. Importantly, following any required strengthening of the deck, retrofitting suction aerofoils is also a simple process, either through welding to the ship's deck or by placing the aerofoil into a restrained standard containerised unit as demonstrated by Econowind with the Ventifoil [9].

2. OPERATION OF SUCTION AEROFOILS

Suction Aerofoils operate on the same principle as a conventional sail, generating lift as air flows over the surface, a component of which can be used to drive the ship forwards. However geometrically, they represent an upright aerofoil with high relative thickness and a movable fin; the cross section modelled in this study is shown in Figure 1 at an example angle of attack (AoA) of 12°.

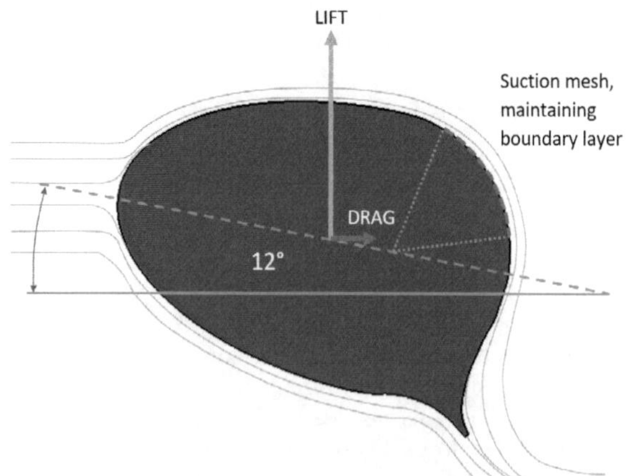

Figure 1: Cross Section of Modelled Suction Aerofoil

Normally, such a thick profile would create a strong adverse pressure gradient in the boundary layer, downstream of where the aerofoil is at its maximum thickness. The adverse pressure gradient would cause the air in the boundary layer to slow down, lose momentum and separate from the surface of the aerofoil (stall). However, by applying suction to the leeward edge, the slow-moving air is removed, leaving a thin fast-moving boundary layer which remains attached to the aerofoil surface much further downstream to create high lift coefficients.

Suction is generated by a fan at one end of the Suction Aerofoil, drawing air through a fine mesh of holes on the leeward side of the aerofoil. Whilst the mesh of holes is present on both sides, the fin may rotate through a limited angle in order to cover the holes on the windward side, leaving suction on just the leeward side. The ability of the fin to switch sides enables the Suction Aerofoil to 'tack', depending on which side of the ship the apparent wind is coming from.

4. COMPUTATIONAL APPROACH

The cross section modelled in this study was symmetrical either side of the chord (except for the fin) and was generated from equations (1) and (2) in a peer reviewed study by Hcini et al [10]. The resulting 10.3m high Suction Aerofoil is shown in Figure 2a, with Figure 2b showing Jacques Cousteau's Turbosail for a visual comparison. The 3D model had a constant cross section with a relative thickness of 66% and a chord of 1.76m.

Figure 2a (left): Modelled Suction Aerofoil
Figure 2b (right): Turbosail Fitted to Alcyone Ship [11]

The geometry in Figure 2a was imported into Numeca FINE/Marine CFD software for the studies, with the Suction Aerofoil modelled in a freestream velocity of 19kts, corresponding to a sea state 4 wind speed ($Re = 1.2 \times 10^6$). A domain 30x30x20m was used, with all external boundary conditions set to far field velocities, except for the outlet which was assigned a prescribed pressure. On the Suction Aerofoil body, wall functions were used with $y^+ \approx 30$.

To resolve the CFD model, the RANS-based Shear Stress Transport (SST) turbulence model was selected, and a mesh convergence study carried out to verify the computational model. The mesh convergence study was carried out for a Suction Aerofoil in free space, with no end plate or ship deck modelled, and no suction applied, therefore a low AoA of 6° was used for the verification.

Figure 3 shows the results of the convergence study for an increasing number of refinements to the mesh. The refinements were applied in boxed regions to create a series of structured meshes, therefore Figure 3 does not show a smooth curve. However, as the refinements were applied to the regions immediately surrounding the Suction Aerofoil and turbulent wake, three meshes were resolved all showing a C_L within 3% of each other, giving confidence in the results.

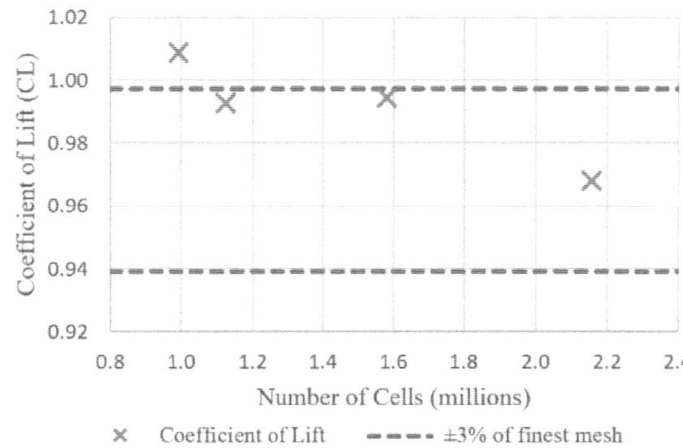

Figure 3: Mesh Convergence for Number of Cells

Carrying the finest mesh forwards with 2.16 million cells, Figure 4 shows the convergence with the number of iterations.

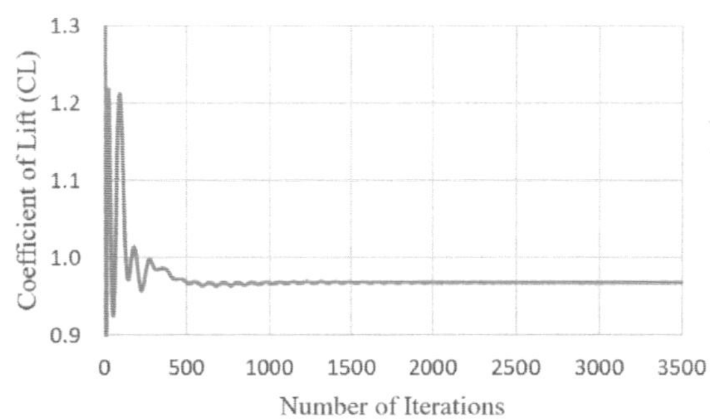

Figure 4: Convergence with Number of Iterations for 2.16 Million Cell Mesh

With confidence in the final mesh (shown in Figure 5), the effect of effects of suction, a ship deck and endplates were investigated in the following sections.

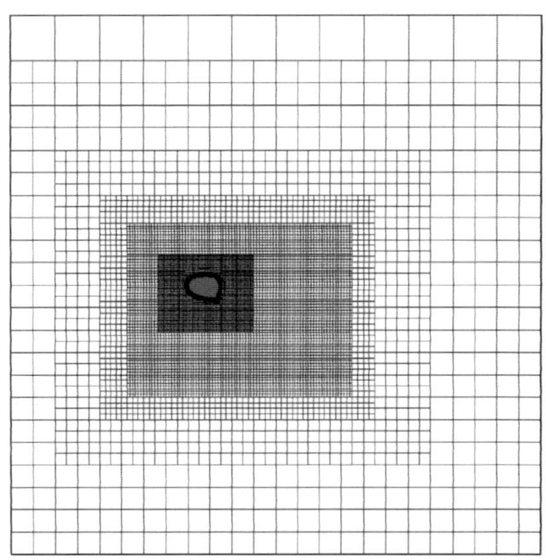

Figure 5: A Cut Through the 2.16 Million Cell Mesh Showing the Suction Aerofoil and Box Refinements

5. EFFECT OF SUCTION

Computationally, the fine mesh of suction holes on the leeward side of the aerofoil was both challenging and time consuming to model. Therefore the suction mesh was simplified to the series of 10mm wide vertical strips (shown in Figure 6) which ran the full height of the aerofoil.

Figure 6: Suction Strips Highlighted Green

The suction strips in Figure 6 were assigned a velocity perpendicular to the surface of the aerofoil, corresponding to a suction coefficient (Cq) as defined by the following equations [9]:

$$Cq = Q / (U_\infty \cdot c \cdot Z_{max})$$

$$\therefore Cq = (U_{suc} \cdot n \cdot dx) / (U_\infty \cdot c)$$

As the suction strips were just 10mm wide, the number of mesh refinements was increased on the surrounding surface of the Suction Aerofoil. Figure 7 shows the result of applying suction for an AoA of 6°; the resulting C_L and C_D were 3.3x and 2x larger respectively in Figure 7b than for the result shown in Figure 7a.

To give an indication of the air flows, for a Cq of 0.023 with the geometry modelled in this study, the velocity of air through the suction strips was roughly equal to the free stream velocity of 19kts – a mass flow rate of 5kg/s into the Suction Aerofoil.

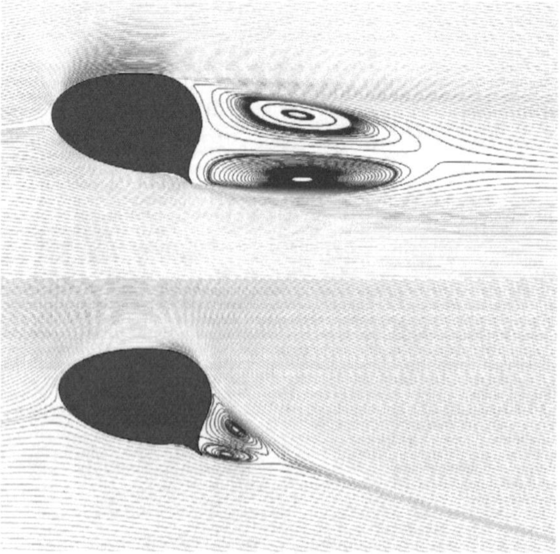

Figure 7a (top): Flow over Aerofoil with $Cq = 0.000$
Figure 7b (bottom): Flow over Aerofoil with $Cq = 0.023$

6. EFFECT OF FLAT DECK

Rather than model the Suction Aerofoil in free space, the aerofoil was modelled on a horizontal flat plate to better represent operating conditions on the deck of a ship. The presence of the simplified deck greatly reduced air spilling from the high pressure windward side to the low pressure leeward side at the base of the Suction Aerofoil, which increased C_L by 16% and decreased C_D by 30% for the 6° angle of attack case.

Figure 8 shows a comparison of the drag forces by section along the height of the suction aerofoil. Notably at the base of the aerofoil ($Z \approx 0$m), the drag forces were much lower when the aerofoil was modelled on a flat deck.

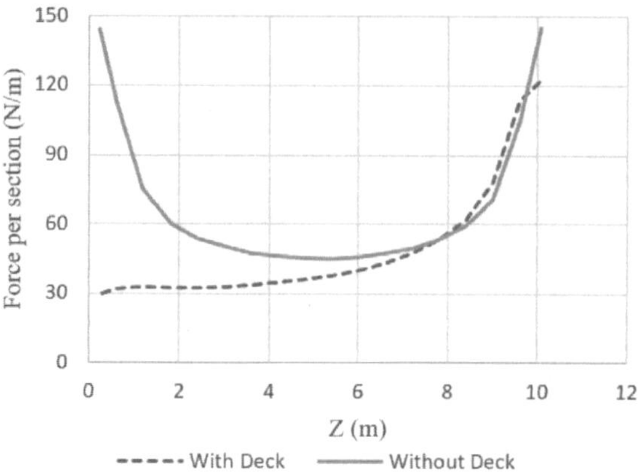

Figure 8: Drag Forces Along Height of Suction Aerofoil

7. RESULTS

With a ship deck added to the model, and at a constant sea state 4 wind speed, the coefficients of lift and drag were solved in Numeca FINE/Marine for varying suction coefficients and angles of attack. The results in this section, plotted in Figure 9, show coefficients of lift for the basic suction aerofoil without an endplate.

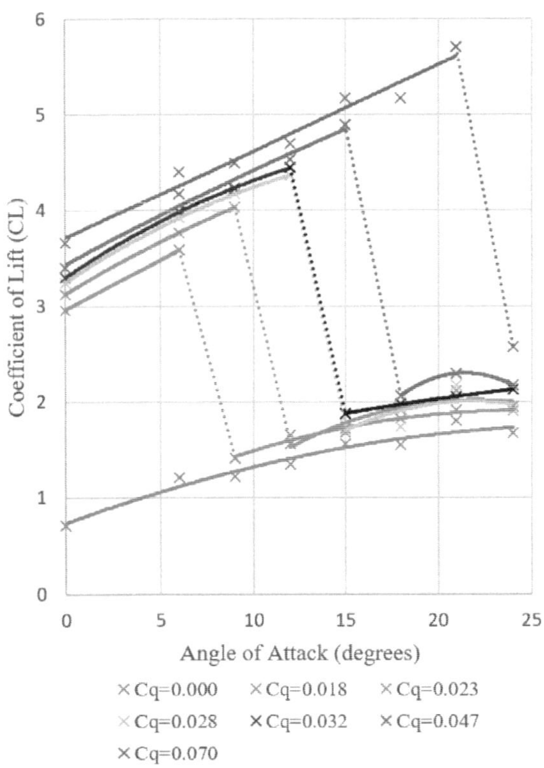

Figure 9: C_L for Varying Cq and AoA

Lift coefficients between 3.0 and 5.7 were achieved where the suction enabled sufficient momentum in the boundary layer to be maintained, preventing stall. In contrast, lift coefficients were restricted to a range between 0.7 and 2.6 when the flow of the suction aerofoil was stalled due to a high angle of attack.

The discontinuous transition between the stalled and non-stalled state was marked by the dashed lines on Figure 9. Notably, the key benefit of increasing the suction coefficient was the increase in angle of attack at which this transition occurred (the critical stall angle), enabling the higher values of C_L to be reached.

Typically, small non-linearities would be expected in the lift coefficients of aerofoils around the critical stall angle, as the C_L-AoA curve flattens off. This was not displayed in the results, therefore given further opportunity to develop the project, more angles of attack would be simulated in Numeca FINE/Marine to investigate this behaviour. However, the curves in Figure 9 do show good correlation to published experimental results from Jacques Cousteau's Turbosail [7]. Both sets of data show that increasing the suction coefficient increases the critical stall angle, and for the non-stalled data ($C_L > 3$), there was an average 18% difference between the results in Figure 9 and the Cousteau's original data [7], despite significant differences in the aerofoil geometry and the varying computational/experimental approaches.

Figure 10 shows the equivalent plot of drag coefficient against angle of attack, showing a general increase in C_D at higher angles of attack and suction coefficients.

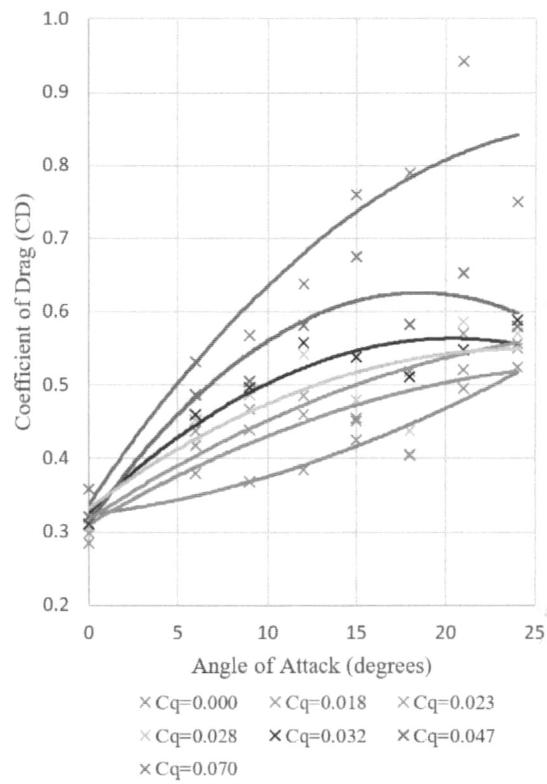

Figure 10: C_D for Varying Cq and AoA

To establish an optimal relationship in the results, contours of C_L were plotted on axes of Cq and AoA in Figure 11. A smoothed curve of best fit was marked (purple) between the points where there was maximum C_L for minimum Cq at any given AoA.

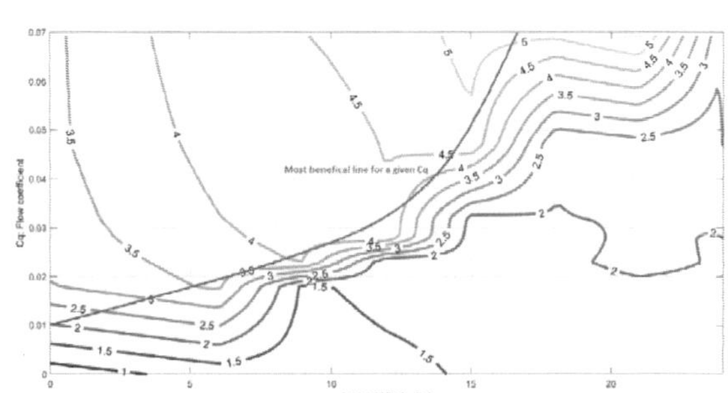

Figure 11: Maximum C_L for Minimum Cq

8. EFFECT OF ENDPLATE

Two further studies were carried out to supplement the results in Section 7 – the first of which studied the effect of an endplate. Figure 12 shows the vortices on the top of a Suction Aerofoil without an endplate, as air from the high pressure windward side spilt over to the low pressure leeward side (an effect well documented and understood in aircraft, where end tip vortices are mitigated by winglets).

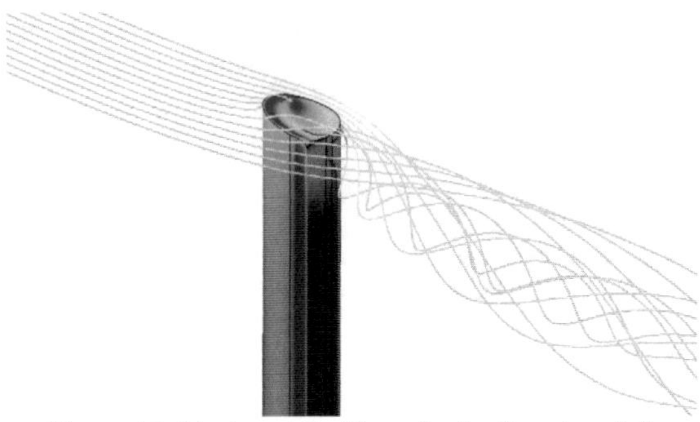

Figure 12: Vortex at the Top of a Suction Aerofoil Without an Endplate

Including the endplate in the CFD model – such as the one shown in Figure 2a – reduced the intensity of the end vortices, resulting in up to ~20% increases of lift to drag ratios for high values of *Cq*, as shown in Figure 13.

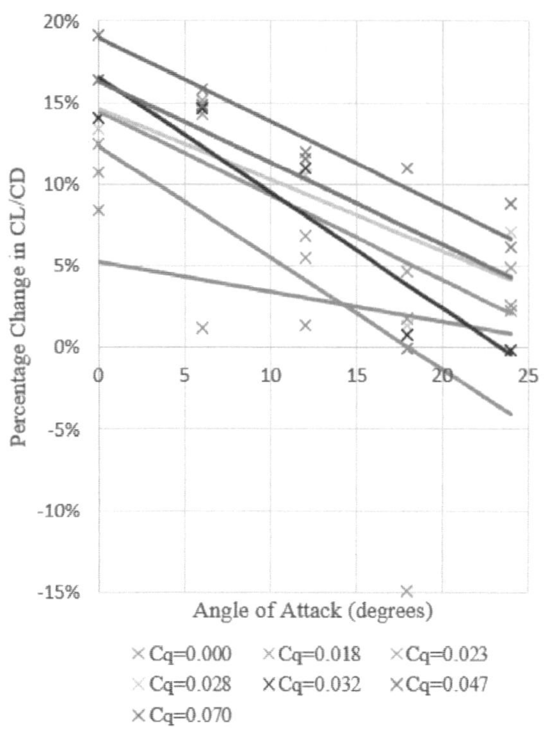

Figure 13: Percentage change in C_L / C_D due to addition of endplate to model

9. EFFECT OF SHIP SIDE

The second supplementary study investigated the effect a side of a ship might have on the wind speed impacting the Suction Aerofoil. For this study, the 10.3m high Suction Aerofoil was placed on the centreline of a simplified ship with beam of 32m and freeboard of 5m, although it should be noted any end affects due to the bow or other ship superstructures were not modelled.

Similarly to the previous studies presented in this paper, a 19kt apparent wind speed was used at an angle of 90° to the ship's heading. However, no suction was modelled on the aerofoil (Cq = 0), with the focus instead on the change in wind speed and direction as the air was forced up and over the ship side.

The results in Figure 14 show that for this particular geometry, the lower ~2.5m of the Suction Aerofoil sees significant reduction in horizontal freestream velocity due to large vortices generated at deck level – as a result, it is expected the nominal effectiveness of the Suction Aerofoil would be affected. In contrast, the remaining upper 7.8m section of the aerofoil saw little change to the wind speed, with just a small increase of ~5%.

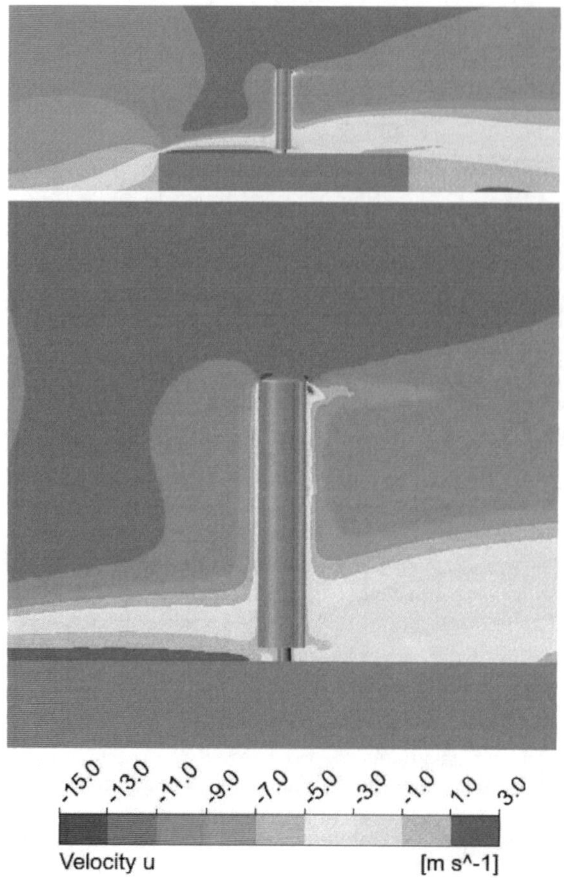

Figure 14: Velocity Contours Across Suction Aerofoil on a Simplified Ship Deck

Despite the wind being forced up and over the side of the ship, Figure 15 shows that there was no significant change to the angle of wind incident on the Suction Aerofoil for this particular geometry. It is important to note however that this simplified model does not account for any complex ship superstructures, different apparent wind angles or any pitching/rolling motion of the ship.

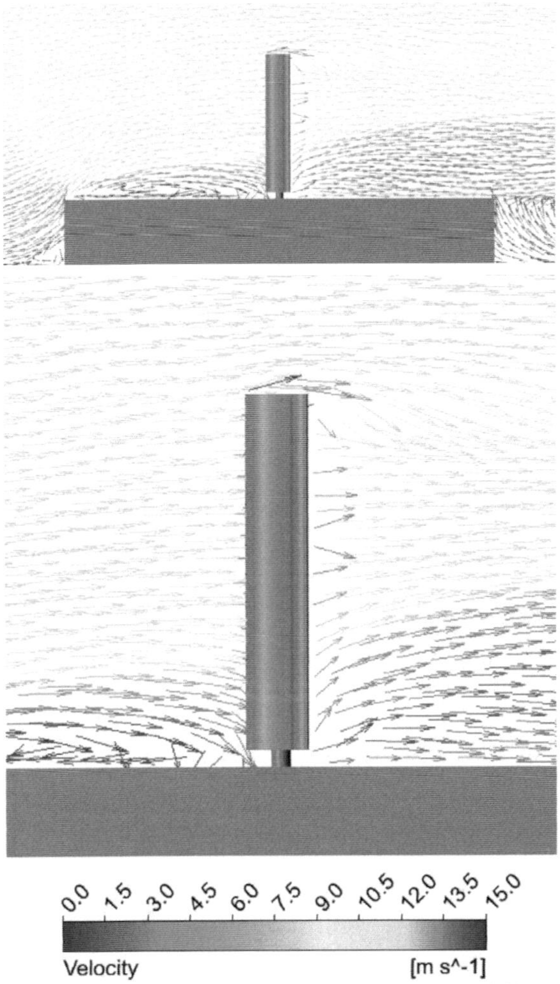

Figure 15: Velocity Vectors Across Suction Aerofoil on a Simplified Ship Deck

9. CONCLUSIONS

In conclusion, the Suction Aerofoil studied in this paper was modelled in Numeca FINE/Marine, and a mesh convergence study carried out to verify the subsequent lift and drag coefficients presented in Figures 9 and 10. These results were also validated against the experimental results of Jacques Cousteau [7] giving further confidence in the modelling.

The plots of C_L and C_D showed that application of suction to the leeward side of the aerofoil increases the angle of attack at which stall occurs, enabling lift coefficients up to ~5.7 to be achieved with the maximum suction coefficient of 0.070 studied in this paper.

Furthermore, by adding an endplate to the top of the suction aerofoil, the lift to drag ratio could be improved by up to ~20% at low angles of attack. This is particularly important if the apparent wind is coming from the forward quarter/bow of the ship, where only a small component of the lift generated by the suction aerofoil would contribute to forward thrust, and the drag generated may contribute to the sideways slippage of the ship, or even act towards the stern and oppose the ships forward motion.

It was recognised however that the ship's freeboard would impact the nominal effectiveness of the Suction Aerofoil, with the wind speed greatly reduced on the lowest 2.5m of the aerofoil for the geometry shown in Figure 14. Therefore there is clearly potential for future work building on this study, firstly to determine C_L and C_D values at varying AoA and Cq with the ship side model, and then also to account for more complex ship superstructures, the bow, different freeboard heights, raising the aerofoil further above the deck, or any pitching/rolling motion of the ship.

Even with the significant potential for future work however, the results in this study already show Suction Aerofoils are capable of producing the high lift coefficients required to provide auxiliary propulsion to a commercial ship, and that they are just one of the promising wind assisted propulsion technologies which should all be considered essential in the efforts to reduce maritime engine exhaust emissions.

10. ACKNOWLEDGEMENTS

The authors wish to thank Jake Rigby for supporting this research at BMT; also Beth Storey, Cameron Barry, Sam Sparks, Shona Cunningham and Tom Beard for their advice and contributions to the project.

It is BMT's intention to claim copyright for this work. The kind permission and resources granted to the authors by BMT are acknowledged with thanks. All findings, ideas, opinions and errors herein are those of the authors and are not necessarily those of BMT Defence and Security UK Ltd.

11. REFERENCES

[1] BMT-led VTAS Project, available online at: *https://vtas-fes.com/*

[2] International Maritime Organization, 'Key Findings', *Third IMO GHG Study*, 2014.

[3] International Maritime Organization, 'Levels of Ambition', *IMO Initial Strategy on Reduction of GHG Emissions from Ships*, 2018.

[4] DNV GL, 'EEXI – Energy Efficiency Existing Ship Index', 2021, available online at: *https://www.dnv.com/maritime/insights/topics/eexi/index.html*

[5] IMarEST North East Coast Joint Branch, 'Webinar: IMO Greenhouse Gas Targets for 2050', Dec 2020, available online at: *https://www.imarest.org/events/category/categories/branch-events/uk/north-east-coast-joint-branch/webinar-imo-greenhouse-gas-targets-for-2050*

[6] Buckingham, J. & Pearson, D., 'Modelling Alternative Propulsion Technologies for Merchant Vessels', *Power & Propulsion Alternatives for Ships*, Jan 2019.

[7] International Windship Association, 'Listening to the Wind: Interview with BAR Technologies', Mar 2021, available online at: *https://www.youtube.com/watch?v=63Ae2MYs4es*

[8] Charrier, B. et al, 'Foundation Cousteau and Windship Propulsion 1980-1985 System Cousteau – Pechiney', *Journal of Wind Engineering and Industrial Aerodynamics, Vol. 20, Iss. 1-3, pp. 39-60,* July 1985

[9] Econowind, available online at: *https://www.econowind.nl/*

[10] Hcini, C. et al, 'A Turbosail Profile Analysis Code Based on the Panel Method', *Energy, Vol. 118, pp. 147-155*, Jan 2017

[11] Hernández, M., 'Alcyone', available online at: *https://www.marinetraffic.com/en/ais/details/ships/shipid:171392*

12. AUTHORS BIOGRAPHY

Will Hopes is a final year undergraduate at the University of Bath studying towards an integrated masters in Mechanical Engineering. He previously spent a year on placement (June 2019 to July 2020) with BMT Defence & Security UK Ltd and has a keen interest in marine engineering and sustainable technologies. Whilst studying for his degree, Will is also proud to have been awarded undergraduate scholarships from both the IMechE and UKNEST, which have helped him pursue his interests in engineering.

David Pearson CEng MIMechE is a Principal Marine Engineer at Steller Systems Ltd, formerly of BMT Defence & Security UK Limited, Bath, UK. David was a Technical Specialist for the Vessel Technology Advice and Support project for the ETI. As part of his consulting work in military and commercial vessel design, David has modelled zero-carbon transport technologies for a variety of projects and has published several papers and articles describing the work and continues to explore the potential of new energy saving technologies.

John Buckingham CEng FIMechE is the Chief Mechanical Engineer of BMT Defence & Security UK Limited, Bath, UK. John was the Chief Technologist for the ETI-funded Vessel Technology Assessment System project. John leads studies into marine power and propulsion, air-conditioning and hydraulics within the company and has published several papers on these subjects. Since 1999, John has been involved in a range of studies to assess the utility of energy saving technologies for the commercial and military marine.

EXPERIENCES OF ROTOR SAIL INSTALLATIONS ON VARIOUS SHIP TYPES AND THE EMISSION REDUCTION POTENTIAL OF ROTOR SAILS

J Kuuskoski, Norsepower Ltd, Finland

SUMMARY

Norsepower's Rotor Sails have been installed on a variety of ship types since 2014. This paper describes the technical and operational requirements for Rotor Sail installations on tankers, ro-ro-vessels, ferries and bulkers. Third-party verified performance results are presented as well as customer feedback on how the Rotor Sails are viewed by the ship's crew. Operational requirements such as air draft restrictions and cargo operations can be complied with by applying a tilting Rotor Sail. Aspects related to tilting Rotor Sails and the first experiences from installations are described.
The global shipping community faces enormous challenges in finding feasible and proven solutions to reduce their environmental impact and specifically fossil fuel related emissions. This paper discusses wind propulsion technology's potential to be a significant part of the emission reduction solution in today's ship designs.

1. INTRODUCTION

1.1 MODERN WIND PROPULSION

Wind propulsion, in the form of sails, was the only propulsive system for large ships moving goods and people for hundreds of years, before the development of engines and fossil fuels led to implementing more powerful propulsion solutions for ships. As a result, the speed of ships increased significantly, driving ship designers and shipowners to compete for customers with faster delivery times and, most of all, reliable sailing schedules. Sails faded out to remain only as a very niche solution for leisure ships and boats.
The existing and foreseeable environmental requirements and energy efficiency requirements for the commercial shipping fleet, particularly through the regulation determined by the International Maritime Organization (IMO) for CO_2 emission reductions of 40% to be achieved by 2030, have put wind propulsion solutions back on the radar for ship designers. All new ship designs must increasingly factor in compliance with multiple environmental indexes and regulatory requirements that dictate how fast a ship is allowed to sail or what type of technologies should be applied. Some people might consider this to be a frustratingly complex development but, when looking at the big picture and the need to reach zero carbon goals, it's inevitably the only way forward for the maritime and shipping community.
Norsepower's solution for a modern wind propulsion system is the Rotor Sail which is a modernised Flettner rotor. Rotor Sails are an efficient way to ensure that ships will be competitive with today's typical bunker fuels as well as tomorrow's green fuels.

1.2 ROTOR SAIL TECHNOLOGY

Rotor Sails are a modernised version of Flettner rotors, mechanical sails which utilise a principle called the Magnus effect. The basic principle is that a rotating object induces a pressure difference between its sides when it is exposed to a flow (in this instance, wind). Essentially, Flettner rotors are lift producing devices like airplane wings or traditional sails of a sailboat. However, The Norsepower Rotor Sail is around ten times more efficient than a conventional sail, because more lift is produced using a much smaller sail area. When wind conditions are favourable, the Rotor Sails allow the main engines to be throttled back, saving fuel and reducing emissions while providing the power needed to maintain speed and voyage time. This makes Rotor Sails a very effective source of auxiliary wind propulsion on ships.
Norsepower's successful modernisation of the Flettner rotor has been proven in several reference projects. The positive results of third-party verified performance measurement campaigns, as well as feedback from the shipowners and crew, have been the catalyst for Norsepower to develop five standardised models to fulfil the needs of a variety of ship designs.

2. CUSTOMER BENEFIT – REDUCING EMISSIONS AND FUEL CONSUMPTION

2.1 PERFORMANCE MEASUREMENT CAMPAIGNS

All new technologies must go through a market acceptance process. Such processes can be lengthy if the customer benefits are not clearly proven and openly presented. Norsepower's target is to accelerate market acceptance by providing results of the third-party verified performance measurement campaigns.
The accurate measurement of the thrust force and related propulsive effect for a sail is not straight-forward. Therefore, specialist companies performed the measurements and analyse the results to ensure precise conclusions. The following table 1. is a summary of the performance measurement campaign results on the first three ships installed with Rotor Sail by Norsepower.

	Estraden	Viking Grace	Maersk Pelican
Measurement party	NAPA	ABB, Chalmers, NAPA	LR
Measurement period	Apr 2015 - Dec 2015	Jul 2018 – Mar 2019	Sept 2018 – Aug 2019
Rotor Sail configuration	Two 18x3*	One 24x4*	Two 30x5*
Average savings [kW]	450	207 - 287	8.2%
Annual fuel savings [ton]	441	230 - 320	8.2%
Annual CO_2 savings [ton]	1380	630 - 890	1400

Table 1. Rotor Sail performance measurement results.

Different measurement methods were used by third-party specialist performance monitoring companies to assess the thrust performance of the Rotor Sails. The results of these full-scale measurement campaigns fell closely in line with Norsepower's own performance simulator, which was used to estimate expected performance before installation of the systems started.

2.2 WIND RESOURCES AROUND THE GLOBE

Wind conditions vary in different ocean areas and regions. Global wind statistics are available to extract information for simulating the potential of wind propulsion in various sailing routes and regions. Figure 2 presents the average wind speed variations in different regions [1].

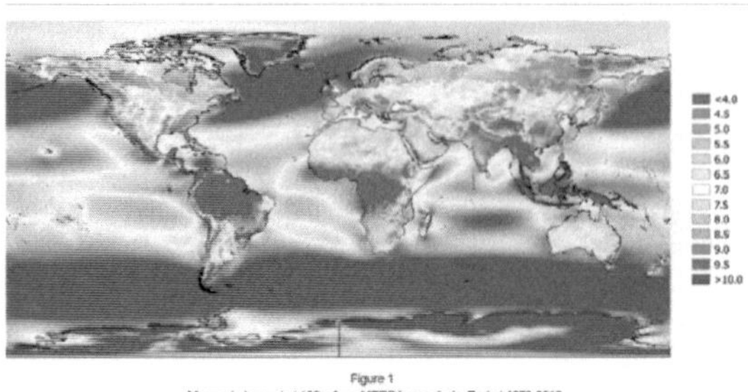

Figure 2. Mean wind speed at 100 m (period 1979-2013) [1].

Norsepower has developed a simulation programme which utilises the Global Wind Atlas database for estimating the thrust performance of Rotor Sails on different routes. Simulation results reveal alignment with third-party measurement campaign results described in the previous chapter. Simulation case studies are performed routinely for new customer projects to gain an understanding of the thrust performance and resulting emissions and fuel savings to be expected during the ship's operations. Tens of simulations were performed for one model 30x5m Rotor Sail installed on a ship sailing at 15 kn to create the map in Figure 3. The colours highlight the varying propulsion power equivalent thrust performance in different regions. The yellow and orange routes are most advantageous in terms of achieving the most significant fuel savings and the blue routes provide less saving potential.

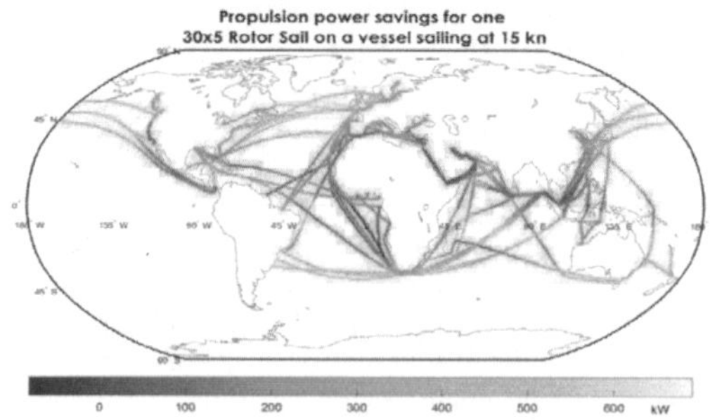

Figure 3. Thrust performance expressed in propulsion power equivalent kW's.

3. OPERATIONAL EXPERIENCES AND FEEDBACK FROM A RECENT DELIVERY PROJECT

3.1 SEA-CARGO'S SC CONNECTOR

Two model 35x5m Rotor Sails were retrofitted on the ro-ro vessel *SC Connector* in late 2020 (Fig.4). The ship is owned by the Bergen, Norway-based shipowner, Sea-Cargo. The ship's main sailing area is the North Sea. Wind conditions in the region are favourable and the potential for reducing the vessel's fuel consumption and related emissions is expected to be around 25%.
The ship passes under several bridges on its regular operating routes, therefore a maximum allowed air draft limited the height of Rotor Sails with a fixed ie. non-tilting foundation. The first tilting Rotor Sail model was implemented to enable installation of the largest Rotor Sails to maximise the opportunity for the highest thrust performance. The tilting foundation module includes a hydraulically operated tilting system able to tilt the Rotor Sails to almost horizontal while the ship sails in a matter of minutes (Fig. 5).

Figure 4. SC Connector with two model 35x5 Rotor Sails.

Figure 5. SC Connector with Rotor Sails tilted.

Customer feedback and description of operating experiences confirmed the expected performance levels of the Rotor Sail, although current results are more qualitative as further detailed measurement campaigns need to be performed.

Statements from the customer include:
"During testing, the two Rotor Sails on the SC Connector have generated a maximum thrust force equivalent of 7 MW propeller shaft power."

"10 tons of fuel were saved during one North Sea passing from Norway. All in all, the expected annual savings are around 25%"

"In perfect wind conditions the vessel will maintain regular service speed by sail alone."

"The crew of SC Connector has noted that the Rotor Sails have made the seakeeping ability of the ship considerably better by reducing the roll and by enabling higher speed in harsh weather."

Such statements demonstrate that Rotor Sails are an efficient, robust and reliable solution to reduce emissions and fuel consumption.

It's worth noting that the last comment refers to the ship's ability to maintain a higher speed in harsh weather conditions due to reduced rolling. Such experiences and benefits can only be realised in the day-to-day operations of the ship.

4. ROLE OF ROTOR SAILS IN DECARBONISING GLOBAL SHIPPING

Norsepower estimates that Rotor Sail technology is technically applicable to approximately 30,000 vessels in the current global fleet of ships larger than 500 GT. The main requirement for suitability is available deck space for installation of the foundation. Other factors to consider are exposure to wind flow and availability of electrical power for the Rotor Sail. Economic feasibility is determined based on the ships typical operating area and operating profile. Generally speaking, payback periods of 3-9 years can be expected with today's bunker prices and without other costs on emissions. This payback figure is estimated to decrease dramatically as new, more costly, low carbon fuels become more prominent in the market.

Total CO_2 emissions by emitted from shipping in 2018 stood at 1.056 billion tons [2]. Norsepower estimates that Rotor Sails could reduce the annual CO_2 emissions by more than 80 million tons if they are installed on the entire available fleet.

Secretary-General of IMO, Mr. Kitack Lim, states in the foreword of the Fourth IMO Greenhouse Gas Study 2020: *"The Study demonstrates that whilst further improvement of the carbon intensity of shipping can be achieved, it will be difficult to achieve IMO's 2050 GHG reduction ambition only through energy-saving technologies and speed reduction of ships. Therefore, under all projected scenarios, in 2050, a large share of the total amount of CO2 reduction will have to come from the use of low-carbon alternative fuels."* [2].

This message reinforces the need to accelerate the transition to alternative fuels and highlights the benefits of energy savings technologies with the increasing costs of both use of fossil fuels and new low- and zero-carbon fuels. Wind is the fuel of nature which is already widely used in power generation and is now available for shipping as well.

5. REFERENCES

[1] The Global Wind Atlas, Final Report by DTU Wind Energy (Oct. 30, 2015).

[2] Fourth IMO Greenhouse Gas Study 2020.

6. AUTHOR BIOGRAPHY

Jukka Kuuskoski has a M.Sc. (Naval Architecture and Marine Engineering) and an eMBA from the Helsinki University of Technology. He worked with ABB's marine business unit for 15 years in management positions of delivery projects, sales and marketing and service operations. During the following 7 years, Jukka worked in the wind power industry with Winwind Oy, a wind turbine manufacturer, and with Element Power, a wind power project developer. He joined Norsepower in 2016 and is responsible for the sales and marketing of Norsepower's Rotor Sails.

SPEED TRIAL VERIFICATION FOR A WIND ASSISTED SHIP

S Werner, **J Nisbet**, **A Hörteborn,** SSPA Sweden AB, Sweden
R Nielsen, Scandlines, Denmark

SUMMARY

As the number of wind assistance installations in commercial shipping grows and the industry matures, the need for full-scale verification of the performance increases. Standard procedures or guidelines for conducting such full-scale trials of are still lacking. One strategy is proposed and discussed here. The method is demonstrated using a speed trial conducted with Scandlines' hybrid ferry Copenhagen equipped with a rotor sail. The trial result is extrapolated to yearly power saving using a statistical route analysis. With this approach, the result can be derived at a feasible cost, within a limited time frame and using commercially available tools and established procedures.

NOMENCLATURE

ρ_{air}	Density of air (kg/m^3)
AWA	Apparent Wind Angle (°)
AWS	Apparent Wind Speed (m/s)
C_L	Lift and drag coefficient (-)
C_D	Drag coefficient (-)
$C_{D\ rotor}$	Drag coefficient of rotor (-)
COG	Course over ground
CFD	Computational Fluid Dynamics
D	Rotor diameter (m)
DOF	Degrees of freedom
EEDI	Energy Efficiency Design Index
H	Rotor height (m)
ITTC	International Towing Tank Conference
Ps	Shaft power (kW)
Pd	Delivered power (kW)
STW	Speed through water (knots)
SOG	Speed over ground (knots)
TWA	True Wind Angle (°)
TWS	True Wind Speed (m/s)
TWS_{10m}	True Wind Speed at 10 m above sea (m/s)
VPP	Velocity prediction program
V_{ref}	Nominal ship's speed (knots)

1. INTRODUCTION

Following IMO's request to reduce green-house gas emissions from shipping [1], wind propulsion has emerged as a feasible solution. The number of cargo vessels equipped with wind assistance technology is still low but rapidly increasing. Flettner rotors are so far the most common type, with several manufacturers active on the market, but there are also other types like rigid wings, soft sails and suction wings.

The emission reduction that a wind assistance technology can offer is typically predicted at design stage using numerical or experimental tools. After installation onboard a vessel, there is for several reasons a need to verify the performance in full-scale. The ship owner may request confirmation of the expected savings. The manufacturers and designers need validation data in order to improve the design and prediction methods for future installations. The shipping community needs confirmed performance indicators to be able to compare different technologies and for fact-based investment decisions. The lack of verifiable data on the fuel savings potential is reported to be one of the key barriers for market uptake of wind assistance technology [2]. Last but not the least, authorities request trustworthy verification of energy reduction for legal indicators such as EEDI [3], and other voluntary notifications of sustainable shipping.

Apart from confirmation of the emission reductions, post-installation testing can include other aspects like safety, manoeuvrability, stability, noise and working conditions. This paper focus entirely on power consumption, which is directly linked to green-house gas emissions. The other aspects will be addressed in future work. The scope is limited to wind assistance, i.e. installations that can reduce a vessels fuel consumption when activated, and when de-activated lets the ship function as a conventional vessel.

Since the wind propulsion industry is fairly new, at least on a larger commercial scale, the community has not converged towards a standard procedure for conducting full scale verification tests. As the industry matures and the number of manufacturers and installations increases, the need for standardised methods for full scale verification grows. The present work aims to contribute to the development and harmonisation of methods for full-scale verification of wind assistance vessels. A methodology is proposed, discussed, and demonstrated for the ferry m/v Copenhagen equipped with a rotor sail shown in Figure 1.

Figure 1. m/v Copenhagen with a Norsepower rotor.

The work is a part of EU Interreg project WASP, running from 2019 to 2023. Further demonstrations for four other vessels are planned within the WASP project. While the primary aim of these five verification campaigns is to confirm the energy saving of the wind propulsion installations, they will also generate experiences and knowledge on the testing methodology and identify advantages and pitfalls of various methods and strategies.

The full-scale verification tests for wind assisted ships that have been published in recent literature can be divided in two different strategies:

- Long-term monitoring. The fuel consumption is logged using the ship's performance monitoring system and periods with and without the wind assistance are compared. This method was used to evaluate two different rotor sail installations: Viking Grace and Maersk Pelican [4].
- Short trials. Dedicated tests over less than a few hours and comparison of speed or power with the device turn on and off. This method is reported briefly for E-Ship 1 [5] and Fehn Pollux [6].

The present study employs the short trial strategy and seeks to re-use as much as possible from the existing and well recognised standards for speed-trials (ISO [7], ITTC [8]), and just modify them where it is needed to fit the purpose.

2. CASE

The RoPax ferry m/v Copenhagen (L=156.45m, B=24.6m, IMO 9587867) operates the route Gedser-Rostock. It is equipped with a 5m x 30m Norsepower rotor sail with end plate. (See ref [4] for a description of the rotor technology.) The rotor is positioned longitudinally around mid-ship, 17.2 m above design water line. The rotation speed of the rotor is set automatically by its control system, based on the measured apparent wind speed. The anemometer is positioned in the top of the signal mast over the bridge. The vessel is driven by two Azimut thrusters and a centre propeller with controllable pitch.

3. CONDUCTION OF SPEED TRIAL

3.1 SETTINGS

On March 6-7, 2021, a speed trial was performed with m/v Copenhagen with the purpose of evaluating the performance of the rotor. The Trial Team present onboard included Ship Master Captain Alan Bach, Scandlines' Naval Architect Rasmus Nielsen, and Sofia Werner, SSPA Sweden AB. The trial was planned and conducted by the Trial Team in cooperation.

The trial was conducted off Gedser in an area of sea water depth 24-25m (Figure 2.). The trial was conducted at night and therefore no visual observations of the wave height could be made. An external weather source (fcoo.dk) reported a wave height of 1 m from west. There is unsignificant tidal current in the area but a constant current from northwest was reported by the external weather provider. The true wind measured with the ship's sensors was 8-9 m/s.

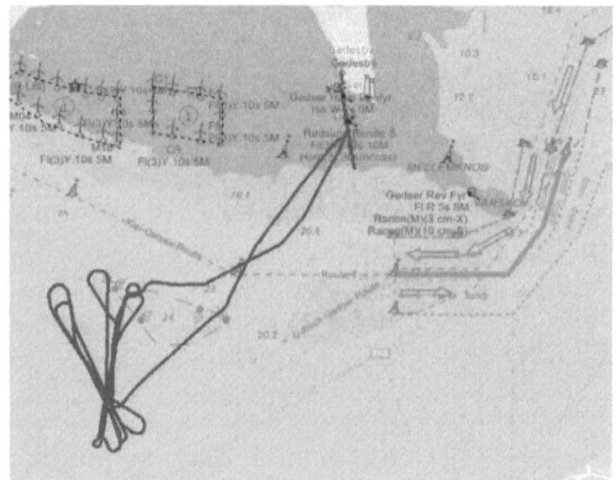

Figure 2. Trial area off Gedser

3.2 DATA ACQUISITION

The speed over ground and track were recorded from the DGPS every 5 seconds. The Azimut power and rpm were recorded every 5 seconds from the electric engine power output. Speed through water from the ship's doppler log, heading from the gyro and the rotor rate of revolution were recorded every 60 seconds. The relative wind speed and direction was measured using the ships anemometer at the mast top with a frequency of 60 seconds. The rotor power consumption was recorded from the electric engine every 30 seconds.

3.3 SPEED TRIAL PROCEDURE

The trial was conducted according to the principles in ISO 15016 / ITTC 7.5-04-01-01.1. The trial program included four double runs according to the plan in Table 1. Each run was 10 minutes long. Constant heading was kept during the runs using the ship's autopilot. The ship's thrusters were set to constant shaft rate. The centre propeller was not engaged during the trial. The rpm of the rotor was set automatically by the rotors control system.

After having reached the trial area, the global wind direction was identified by turning the ship through the wind while reading the anemometer. The direction of the first run was determined to be around 90 degrees off the global wind direction. The first double run was performed without the rotor spinning, directly followed by a double run in the same direction with the rotor turned on. This was then repeated aiming for a true wind direction of 40/-140 degrees. The tracks are shown in Figure 3, where the circles mark the start of a run.

Run	True Wind Angle	Rotor
1	90° port	Off
2	90° s.b.	Off
3	90° port	On
4	90° s.b.	On
5	40° port	Off
6	140° s.b.	Off
7	40° port	On
8	140° s.b.	On

Table 1. Planned trial program. True wind angles relative to ship bow.

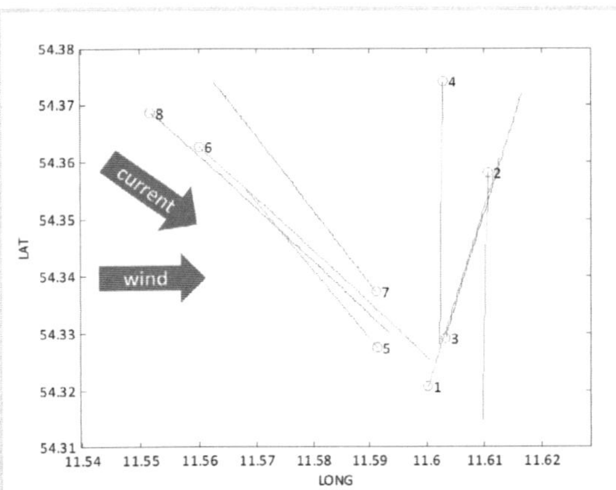

Figure r. Tracks of trial runs. Circles mark the start of each run.

4. SPEED TRIAL ANALYSIS

4.1 CURRENT

One of the key outputs from a speed trial is of course the ship's speed through water (STW). However, speed logs are in general too inaccurate [9, 10]. Measuring the speed over ground (SOG), historically by using land marks and accurate timing, and nowadays using GPS, is much more accurate. The problem is that the SOG readings are affected by current. In the standard ISO/ITTC speed trial procedure [7,8], current is eliminated using double runs, .i.e. two runs in reciprocal direction with the same engine power setting. The correction method, either the so-called Means of Means or the Iterative method (see [11] for further details) effectively compensate for the current, under the condition that the engine power is constant over the two runs, and that the effect of wind and waves can be estimated and subtracted. For the present case, the first condition cannot be fulfilled for other wind directions than exactly +-90 degrees from the bow. For all other wind directions, there is an additional, unknown, propulsive power that is different between the two runs.

The solution applied in this case is to use the ship's log directly, and therefore avoid the need to correct for current. A bias error on the speed log of 0.1 knots was estimated by comparing with the GPS and assuming a slowly changing current, and this was extracted from all runs. However, since the aim of the current trial is to compare the runs with and without rotor, a small bias error in the speed log readings will have no influence on the result.

4.2 DRIFT

According to the standard ISO/ITTC procedures, no correction is applied for drift or rudder angle. It is anyway interesting to examine whether the rotor contributed to any considerable drift. The drift, derived as the difference between the course over ground (COG) and heading, was compared between the runs with and without rotor. The derived drift was between 2 and 5 degrees, but there was no difference between the runs with and without rotor except for the run at true wind angle 40 degrees. The side force from the running rotor then increases the drift angle by 2 degrees. This small increase does not contribute to any measurable increased resistance.

4.3 WIND

The true wind during the trial is derived from the apparent wind measured using the ship's anemometer and the ship's speed. As can be seen from the black lines in Figure 5, the derived true wind appears to change magnitude depending on the ship direction. This is of course unreasonable, and the reason for this is the disturbance of the hull superstructure. As this is a common phenomenon, the standard ISO/ITTC procedures include a strategy for dealing with this error source. The method is denoted "wind averaging" and prescribes that the derived true wind from an up-run and its corresponding down-run are averaged.

The red curve in Figure 4 marks the true wind after averaging between double runs. This corresponds to a true wind speed between 8 and 10 m/s at reference hight 10m above sea level, from V-NV direction. This fits with the externally reported weather.

Table 2 lists the derived true wind after averaging and correcting to 10m height according to ISO standard, as well as the apparent wind computed back based on the averaged true wind and the ship's heading and speed. These are the wind properties that are used in the speed trial analysis.

4.4 WATER TEMPERATURE, DISPLACEMENT AND SUPERSTRUCTURE RESISTANCE

The measured power for each single run is corrected for the resistance of the superstructure based on ISO/ITTC standard procedure. The wind resistance coefficient is the "Ferry/Cruise ship" from the ITTC procedures. Correction for water temperature and a correction of displacement to baseline displacement are done according to the same procedures.

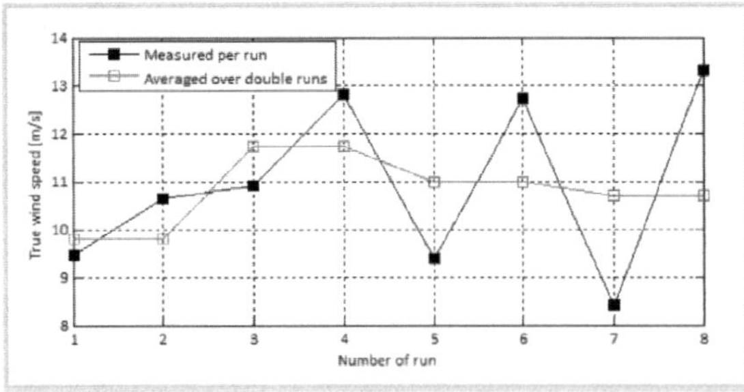

Figure 4. True wind speed, at height of anemometer. The black squares are 10 minutes average of the single runs. The red squares mark the averages between double runs (1&2, 3&4 etc).

Run No		TWS (m/s)		TWA (deg)	
No rotor	with rotor	No rotor	with rotor	No rotor	with rotor
1	3	8.1	9.7	88	78
2	4	8.1	9.7	92	102
5	7	9.1	8.9	41	38
6	8	9.1	8.9	139	142

Run No		AWS (m/s)		AWA (deg)	
No rotor	with rotor	No rotor	with rotor	No rotor	with rotor
1	3	12.6	15.7	51	47
2	4	12.6	13.1	51	61
5	7	17.3	17.4	25	22
6	8	7.3	6.6	90	90

Table 2. Derived wind after averaging over double runs. True wind is corrected to reference level 10 m above sea.

4.5 IDLING ROTOR RESISTANCE

Since the purpose of the exercise is to derive the effect of the rotor compared to the ship without any rotor, the resistance of the idling rotor during the trial must be subtracted from the runs when the rotor was not used. The rotor resistance is estimated as:

$$R_{rotor} = C_{\mathrm{D\,rotor}} \frac{1}{2} \rho_{air} \cdot H \cdot D \cdot AWS_x{}^2 \qquad (1)$$

The resistance coefficient of the idling rotor, $C_{\mathrm{D\,rotor}}$, is estimated to be 0.5 [12]. AWS_x is the apparent wind speed in the ships longitudinal direction at the height of the rotor.

4.6 POWER CORRECTION

The correction of propulsive efficiency due to the added resistance corrections and idling rotor resistance is derived according to the ISO/ITTC standard using the Direct Power Method, see [7,8] for details.

5. SPEED TRIAL RESULTS

Due to the fluctuating wind, drift and current, the wind direction relative to the ship came out differently from the original plan. The resulting range and distribution of wind angles cover well the operational range of the rotor, though.

To extract the effect of the rotor, the single runs with and without rotor at the same wind conditions are compared. Table 3 and Figure 5 present a comparison of the speed and corrected power between the runs with and without rotor. These results reflect that there are several effects of the rotor:

- Increase of speed due to the additional thrust.
- Reduction of power due to off-loading the propellers. Since engine shaft rate is the same for all runs and forward speed is increased due to the rotor, the advance ratio is increased and with that also the propeller efficiency.
- Changed rudder angles and drift due to side force. These effects are included in the speed and power figures but have not been quantified separately.

Run No		TWA	Δ STW	Δ Ps
No rotor	with rotor	(deg)	(knots)	(%)
1	3	79	1.15	-8.0%
2	4	101	1.20	-4.1%
5	7	38	0.58	-0.0%
6	8	142	0.31	-3.2%

Table 3 Difference of speed and corrected power between speed trial runs with and without rotor.

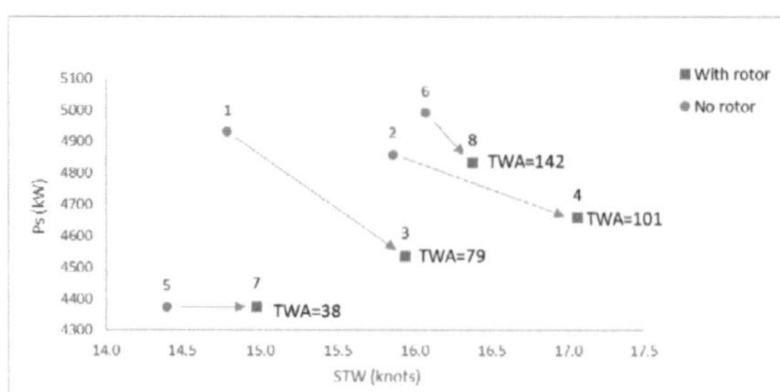

Figure 5. Speed and corrected power from trial

6. GENERALISED ROTOR PERFORMANCE

The result of the speed trial in the previous chapter showed that the rotor contributed to an increased speed as well as a power reduction. In this section, the speed trial result is normalised to derive a power reduction for a given ship speed and reference wind speed. Two alternative methods for the normalisation are used.

6.1 DIRECT NORMALISATION METHOD TO NOMINAL SPEED

To derive a power difference at nominal speed V_{ref}, the power figures from the speed trial analysis is interpolated to V_{ref}, using the shape of the ship's baseline speed-power curve. For the actual vessel, baseline curves have previously been derived by the Ship owner based on speed trials. The interpolation is done by fitting a 3rd order polynomial to the baseline curve and shift it vertically, as Figure 8 indicates.

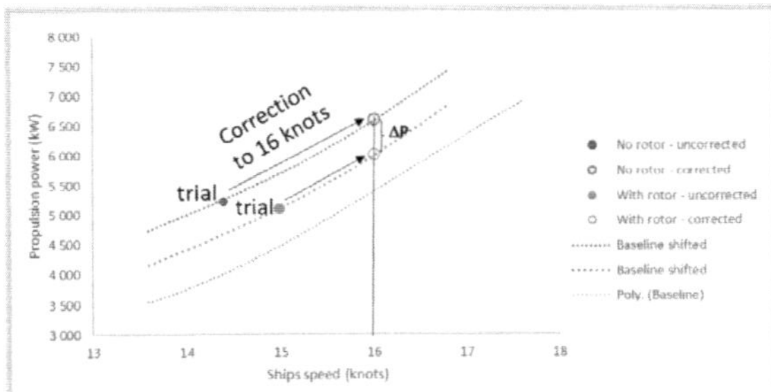

Figure 6. Example of how speed trial result is extrapolated to nominal speed using the shape of the ship's speed-power curve.

The derived power difference is corrected to a nominal wind speed using:

$$\Delta P_{TWS_{ref}} = \Delta P \cdot \frac{TWS_{ref}^{2}}{TWS^{2}} \quad (2)$$

where TWS_{ref} is the reference wind speed and TWS is the true wind speed at the sea trial, at the same height. The reference speed is according to standard praxis given for a hight of 10 m above the sea. The wind variation over height is computed according to ISO 15016 using exponent 1/7. As shown in Figure 7, if the reference speed is 10m/s at 10m hight above sea, the wind is between 10.8 and 12.5 m/s over the rotor. TWS_{ref} is the wind speed at the mid-span of the rotor, which is 11.8m/s in this case.

The resulting power savings are given in Table 4. The "net" numbers at the right-hand side of the table includes the power consumption from spinning the rotor.

Figure 7. Wind profile according to ISO/ITTC [7,8]. (Note that the rotor end plate is missing in the drawing.)

The direct normalisation method includes several simplifications. The translation of a speed increase to a power decrease by shifting the power curves does not fully account for the changed propulsive efficiency. A second simplification is that the changed apparent wind due to a changed ship speed is not included. Therefore, a more advanced method is proposed in the next section.

TWA	ΔPd Gross	ΔPd Net
deg	%	%
78	27	25
102	25	24
38	10	8
142	10	9

Table 4. Direct normalisation method results: Power reduction derived from speed trial and normalized to ship's speed 16 knots and TWS_{10m} 10 m/s. "Gross" means without considering power consumption from rotor, "Net" means with.

6.2 GENERAL PERFORMANCE MODEL

In order to extrapolate the trial results to any arbitrary speed and wind condition, a ship simulation program is used. The program is part of SSPA's inhouse simulation code SEAMAN. The part that is used here is a static 4DOF VPP including propeller and power models. The process includes the following steps:

The starting point is generic lift and drag curves for a rotor of the actual aspect ratio derived using full scale CFD simulations [13]. The aim is to tune the generic rotor curves to the measured speed trial results. To do that, the *rotor thrust force* needs to be extracted from the speed trial results. For that purpose, a ship simulation program is used, which can model the relation between speed, resistance, power and the change in propeller efficiency due to changed speed or propeller load. The model is here based on resistance and propulsion characteristics derived from earlier model test and the result is tuned against the ship's calm water speed-power curve at the actual loading condition, without rotor.

With the help of this speed-power model, it is possible to derive the difference in longitudinal force that match the speed and power differences observed in the sea trial, while considering the effect of changed propulsive efficiency. This difference is assumed to be the rotor thrust force.

The derived rotor thrust forces is non-dimensionalised using the apparent wind speed and air density and compared with the generic rotor values from CFD. For this case, the difference increases quadratically with the apparent wind angle, such that the generic model overpredicts the performance at the larger wind angles. This is in line with what has been observed at SSPA in numerical studies of similar ships. The authors believe that the reason is the effect of the ship superstructure on the incoming flow. In the CFD simulations for the generic case, the rotor is placed on a symmetry plane, i.e without any ship hull. The performance of a Flettner rotor standing on a ship hull has been reported to be quite different from that of a rotor standing on a symmetry plane or on a floor in a wind tunnel. CFD studies by Garneaux [14] showed both increased and decreased performance depending on the wind angle. Vahs [6] reported an increased performance compared to the ideal case for a rotor positioned in the bow of a coaster measured in full-scale. Jones [15] on the other hand, observed decreased performance when placing rotors on tanker ship hull. The discrepancy between the generic rotor and the measured performance can also be due to errors in the CFD simulations. This topic will be addressed further in future work.

The rotor model thrust coefficient is now tuned using the second order polynomial derived from the described comparison. The same correction is applied to the side force, assuming that the ideal rotor C_L/C_D is preserved. This is an assumption, but since side forces is not measured at the speed trial, it is the best possible approach. However, the magnitude of the side force has only a marginal effect on the power gain for the current case.

The force from the rotor can now be derived for any wind condition using the tuned thrust coefficient, air density and the wind speed at the mid-span of the rotor.

When the rotor thrust is negative, it is assumed that the rotor is turned off. In head wind, the rotor will give an added resistance according to equation (1).

The drift and rudder forces are introduced in the ship simulation tool in terms of manoeuvring coefficients based on the bis system model due to Norrbin [16]. In the present case, the manoeuvring coefficients are extracted from SSPA's database of manoeuvring model tests. Added resistance in waves are derived using spectral superposition of response amplitude operators (RAO) (found from model tests in regular waves from SSPA database) and wave spectrum (ITTC) to find mean added resistance in an irregular sea state.

The power saving due to the rotor can now be extracted at any ship's speed and wind condition by executing the ship simulation with the rotor model turn either on or off. Figure 8 show the result at ship's speed 16 knots and TWS_{10}=10m/s, together with the results from the more simplified direct normalisation described in earlier. The two methods coincide within the uncertainty level, which indicates that the simplified method is acceptable. However, this needs to be studied further with other test cases.

Figure 9 presents the power savings at a number of wind conditions. The ship's speed is fixed to 16 knots and density of air is 1.24 kg/m3. The rotational speed of the rotor is dependent on the wind speed and direction according to tabular values provided by Norsepower. The power required to operate the rotor is dependent on the rotational speed and based on information from Norsepower.

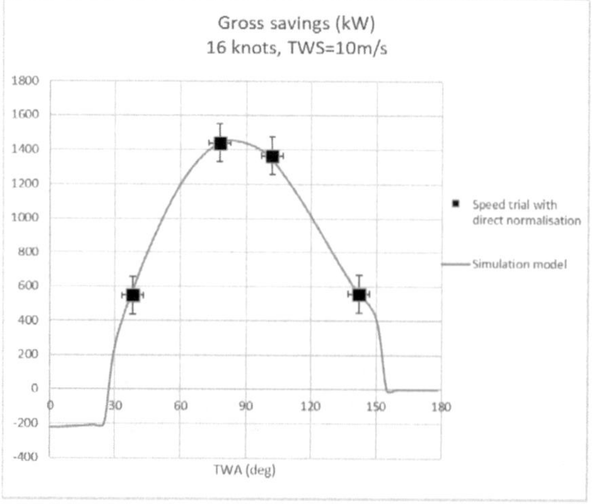

Figure 8. Gross Power saving derived from speed trial with direct normalisation and with ship simulation model tuned with thrust coefficient from speed trial. Error bars as described in section 8.1.

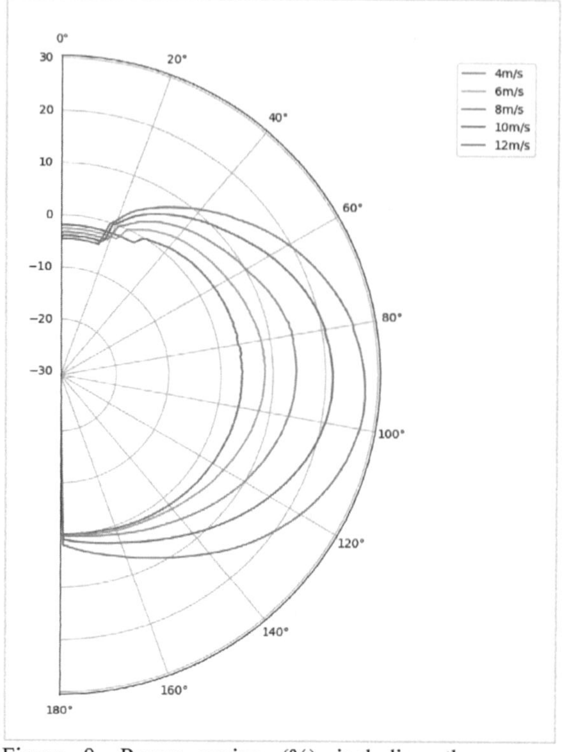

Figure 9. Power saving (%) including the power consumption from spinning the rotor for a variation of true wind speeds and angles. Derived using simulation model tuned to full-scale trials.

7. YEARLY FUEL SAVING

The final step in the analysis is to estimate the early fuel saving. This is done using a route analysis compassing the following parts:

- The ship and rotor model described in the previous section.
- Weather statistics on the route
- A Monte-Carlo based route simulation

7.1 SETTINGS

The route analysis is carried out for the following conditions:

- Fixed speed 16 knots (the ship's service speed)
- Design loading condition
- Air density 1.24 kg/m^3
- Route Gedser and Rostock and back (Figure 10).

Limitations:

- The main engine is assumed to always deliver enough power and torque to reach the intended speed, i.e. no involuntary speed reductions.
- Voluntary speed reductions are not accounted for.
- Hull fouling is not accounted for.

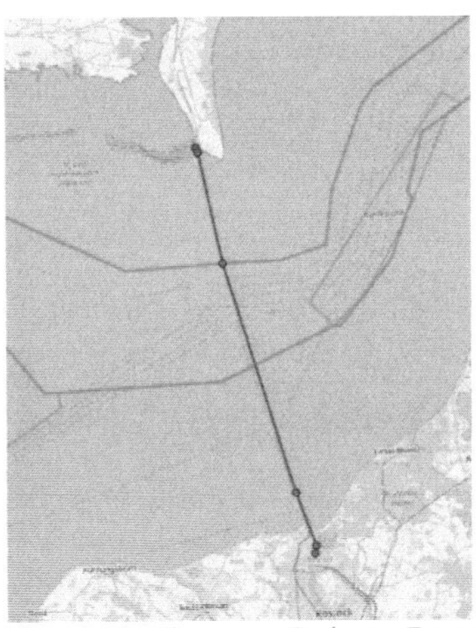

Figure 12. The route Gedser – Rostock

7.2 WIND STATISTICS

Statistical wind distribution for the area is obtained from the Global Wind Atlas [17]. The Global Wind Atlas use the reanalysis ERA5 statistics and apply both mesosacale and microscale modelling in order to get a 250x250m grid of local wind climate. In this case is the weather statistics gathered from a polygon with lower left (11.95175 54.20985) and upper right (12.06024 54.51391) at 10m with a 0.0 reference roughness length.

As a complement, the wind statistics from EEDI Global Weather matrix [2] is also included. It represents the wind on the mayor world-wide trade routes. This wind statistic is not representative for the current ferry route, but it is included here to give a general, route-independent performance evaluation. The comparison of the wind speed distribution in Figure 11 shows that Global Wind Atlas predicts higher wind speeds for the actual sea area than the EEDI global weather matrix.

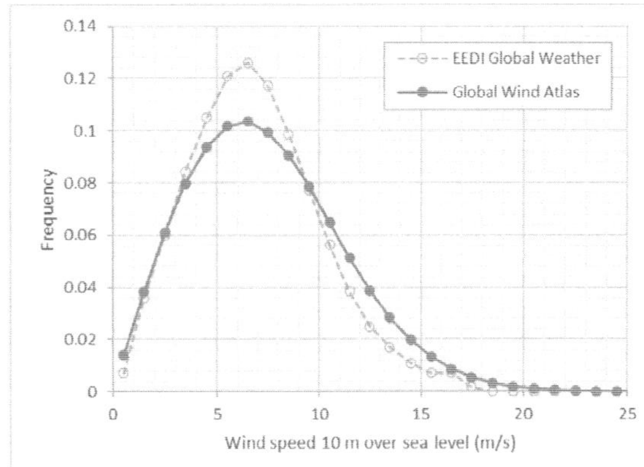

Figure 13. Wind speed distribution from Global Wind Atlas for the actual area, and the EEDI Global Weather matrix [2].

7.3 ROUTE SIMULATION TECHNIQUE

Route simulations are carried out by performing statistical simulations of Monte Carlo type over different combinations of environmental conditions along the route to estimate statistical properties of route energy requirement.

The route between Gedser and Rostock is divided into five legs. For each leg on the route, a discrete joint weather distribution (True wind speeds and True wind angles) is defined. Each leg is treated independently, and leg-wise distributions are assumed to be uncorrelated.

A Monte Carlo simulation is then performed according to the following steps:

1. For each iteration in the Monte Carlo simulation
   a. For each leg on the route
      i. Find the average azimuth and distance that the ship will travel on this leg
      ii. Draw a sample weather condition from the discrete, joint weather distribution. The sample is randomly chosen based on its probability of occurrence, i.e. a weather condition with 2% probability of occurring has a 2% probability of being sampled.
      iii. Evaluate ship performance for this specific weather condition using the performance polar curves from the VPP, linearly interpolating where necessary.
   b. Aggregate the results for the entire route based on the sub-results from each leg.

2. Combine the aggregate route results from all iterations into a single cumulative performance distribution

7.4 ROUTE SIMULATION RESULTS

As the vessel will experience different weather over a period in time (introduced by the Monte Carlo simulations) the power requirement will be different from one journey to another. Figure 12 gives the probability distribution of power for the two cases with and without the rotor employed. It can be noted that for a majority of the trips, the difference is small. The case *with* rotor has, naturally, a larger spread in power demand. The difference is largest at the lower power range (to the left in the figure), which probably convers the days of strong beam wind. This condition is optimal for the rotor, but also not very harmful for the case without rotor. The higher power range (to the right in the figure) covers the days of strong head wind, when both cases struggle.

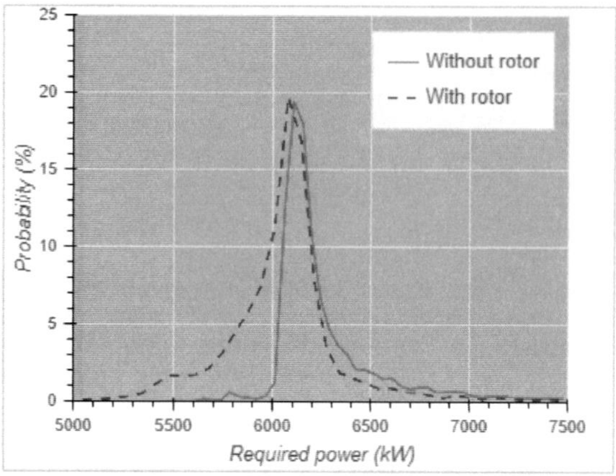

Figure 12. Probability density function for required power on the route to keep 16 knots ship's speed, with and without rotor sail.

Averaged over a long period, the difference in power requirement corresponds to a 3.9% reduction in power, including the power requirement of spinning the rotor. Since the fuel consumption is proportional to the power requirement within the limited design range of shaft rates, the same percentage saving is valid for the expected fuel saving. The corresponding value when using the EEDI Global Weather Matrix as the weather source is 2.0 %.

Weather source	Yearly Power Saving
Global Wind Atlas	3.9%
EEDI Global Weather matrix	2.0%

Table 5. Expected averaged power saving based on different weather sources.

8. UNCERTAINTY ASSESSMENT

8.1 SPEED TRIAL UNCERTAINTY ASSESSMENT

The bias uncertainty of speed trials is stated in the ISO/ITTC [7,8] standard to be 2%. In the present work, the purpose is to derive a power difference, and then the bias error can be assumed to cancel out. The exception is the wind; a bias error of the anemometer will strike differently on the run with rotor compared to the run without rotor.

The precision error of speed trials in general is estimated by Werner [18] and Insel [19] using series of sister ships to be around 7-8%. However, most of this uncertainty is probably due physical differences between sister ships, and trials conducted at different occasions. The precision uncertainty for the current comparative test is probably smaller, but there are no published results to lean on.

Here follows an estimate of the uncertainty of the derived power difference, following ITTC 7.5-02-01-01 (Type A). The authors do not claim it to be a complete uncertainty assessment, but an indication of the magnitude of the larger error sources.

The largest source of uncertainty is the standard deviation of the speed log (see Table 4), which was retrieved at low frequency (1 min). However, comparing with a prediction based on GPS speed shows that the uncertainty is probably less than the standard deviation indicates.

The anemometer also affects the evaluation uncertainty. The fluctuation of the natural wind is high and therefore, higher frequency logging would have been preferred. There is also some disturbance caused by the hull, which is more problematic as it is very difficult to assess. It is difficult to measure the "true" apparent wind hitting the rotor, since all possible locations to place an anemometer is disturbed by the hull or the rotor. On this ship, the anemometer has been corrected using lidar measurements, but the hull disturbance is anyway significant, which the oscillating behaviour of the calculated true wind in Figure 5 indicates.

The experimental design employed in this work implies that single runs with and without rotor *at the same wind conditions* are compared. Possible differences in wind condition between runs that are compared could disturb the comparison. Figure 13a-b show the wind conditions of the runs with and without rotor that are paired for the comparison. It is seen that the conditions are reasonably close within the pairs. Between run 1 and 3 the wind speed increased, which means higher hull windage drag. However, the superstructure resistance is compensated for by a correction of the power (see section 4.4). The air resistance coefficient is taken from the ITTC library and is not ship specific, which could introduce an error in the comparison. The possible error from this approximation is

Variable	Comment, source of uncertainty	Uncertainty of variable (Type A)	Uncertainty of power saving
Heading	Standard deviation of time signal	1 deg	insignificant
STW	Standard deviation of time signal	0.25 kts	240 kW
STW	Comparing with analysis based on SOG	0.1 kts	130 kW
power	Standard deviation of time signal		90 kW
AWA	Standard deviation of time signal Disturbance of hull	5 deg	Secondary effects: hull air resistance, regression of thrust function in Method 2
AWS	Standard deviation of time signal Disturbance of hull Atmospheric boundary layer difference from 1/7 power law	1 m/s	150 kW (on the normalisation to given wind speed)
	Assumptions in the normalisation method. Assessed by varying the input.		50 kW

Table 6. Speed trial uncertainty assessment

conservatively estimated to 10% of the air resistance. However, the wind resistance correction is just up to 3% of the total resistance for run 1-3. This means that the possible error on the power difference is around 0.3%.

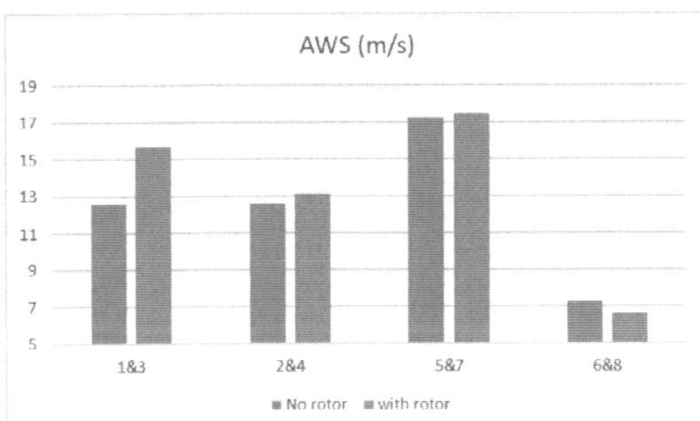

Figure 13a. Apparent wind speed for pairs of runs with and without rotor.

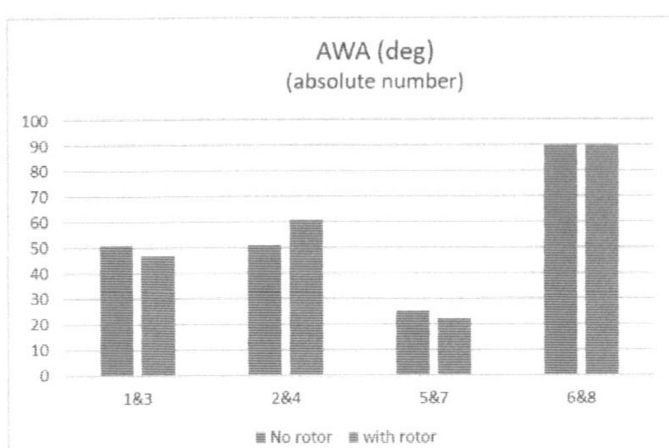

Figure 13b. Apparent wind angle for pairs of runs with and without rotor.

To reduce the uncertainty for coming trials, it is recommended to:

- Use high frequency automatic data collection of speed log
- Use high frequency automatic data collection of anemometer
- Try to correct anemometer for hull disturbance using Lidar or CFD simulations
-

8.2 GENERALISED SHIP MODEL UNCERTAINTY

Simulation models always include assumptions and simplification and cannot mimic the behaviour of complex ship system exactly. This introduces errors in the simulation results.

The comparison between the two normalisation methods in Figure 8 shows that they agree well. The first method is simple and transparent and does not require any speed-power prediction program as the second method does. Therefore, it can be a useful method in praxis.

Both methods rely on assumptions and model test data related to the propulsion factors. The possible uncertainty that this causes could be reduced if the trials were conducted not at constant shaft rate, but instead aiming for similar speed between the runs with and without rotor. This test design will be investigated in future trials.

For the complete generalised model, the manoeuvring coefficients are estimated based on experience and the Azimut thrusters have been modelled as conventional propellers and rudders, since there was no model test of CFD analysis done to extract the manoeuvring coefficient for the actual vessel. This is believed to have insignificant effect on the fuel saving results as the drift was found to be small even for the high wind speeds at the speed trial.

The process of tuning the simulation model to the trial tests is believed to result in an accurate ship model for true wind angles between 35 and 145 degrees from the bow. The resistance that the rotor is assumed to generate in head wind is based on an empirical assumption of resistance of a cylinder. The uncertainty associated with this assumption, in particular the influence of the hull, should be investigated further using numerical tools.

8.3 ROUTE SIMULATION UNCERTAINTY

The weather statistics probably contributes to high uncertainty in the route simulation. The weather provider does not state any uncertainty levels for the data, though. Wind measurements are currently assembled onboard the ship, and this will complement the study later.

The largest uncertainty relates to the actual operation of the vessel and rotor. The annual power saving derived with the route analysis assumes that the rotor is used all the time when the wind conditions allows, i.e. no down-time due to maintenance etc. It is also assumed that the speed is kept constant, i.e. that the crew chose to adjust the engine power to keep the fixed speed when the rotor is in operation, rather than running at a fixed power and "save" time to port. If the latter happens, no fuel saving will be made.

9. METHODOLOGY DISCUSSION

9.1 ASSESSMENT STRATEGY – LONG TERM MONITORING OR SHORT TRIAL

Full-scale verification campaigns for wind assisted ships in recent publications can be divided into two different strategies: Long-term monitoring and Short trials. The selection of strategy does not need to be exclusive; combinations are of course possible. For the present work it was decided to lay the focus on the short trial strategy, and possibly compliment with the long-term monitoring later on. Here follows a short discussion on the reasoning behind this decision.

In the Long-term monitoring strategy, the fuel consumption is logged using the ships performance monitoring system and periods with and without the wind assistance are compared. The advantage of this strategy is that all unpredictable operational aspects are covered: unexpected down-time of wind propulsion device, crew skill in operation, actual experienced environmental conditions. The disadvantages are:

- The uncertainty and scatter of in-service logging data is in general very high. Very long time periods are needed in order to get reliable trends. This means that within the course of the test period with and without the wind assistance technology (e.g. before and after installation), the fuel consumption may be affected by other factors, for example hull fouling, docking, hull cleaning, engine maintenance. There is a risk that this influences on the conclusions.
- The issue above can be avoided by instead deliberately in-activate the wind assistance device for test periods between periods of active use. In that case the ship operator misses out a large potion of potential fuel saving!
- For ships that do not operate on a fixed and regular trade, the logging periods with and without wind assistance will never be similar. Loading condition, wind, waves, current, temperature and so on may heavily disturb the comparison even when to corrected for.
- When post-processing the large amount of data, different filtering and correcting settings may lead to different results, giving ambiguous conclusions.
- The result is valid for the particular trade, and the particular time only. Hence it is difficult to derive general performance indicators that can be compared with other wind assistance installations. This may be irrelevant for the individual ship owner but is a drawback for the shipping community and the further development and uptake of wind propulsion.

The Short trial strategy overcomes the issues above by performing the comparison in unchanged ship and environmental condition. The main advantage is that it safely provides transparent results and general performance indicators that are valid for any route and weather condition. The drawback is instead that the test results is limited to a few conditions. Route simulations and weather statics need to be used to derive an expected fuel saving. Unexpected issues in the operation will not be accounted for. Therefore, this kind of evaluation should be accompanied with long-term monitoring and independent evaluation of the actual operationality.

9.2 SPEED TRIAL METHODOLOGY

The proposed method seeks to follow the standard speed trial procedures as much as possible. The standard procedures prescribe the use of reciprocal double runs as a mean to correct for current so that the ship's speed can be measured using a GPS instead of the ship's log, which is usually inaccurate. The standard correction methods cannot be applied in case of wind propulsion, unless the wind angle is +-90 degrees from the bow.

The here proposed method suggests using the speed log for the ships speed and thereby avoid the need for current correction. Since the objective is to derive a difference between two conditions, with and without wind power device, a bias error of the log will cancel out. Precision error of the log will affect the uncertainty of the evaluation, though. However, the authors believe that it is better to accept this uncertainty, than not to measure at all. The alternative solution to evaluate the wind propulsion device at 90 degrees only is in the authors opinion worse. The hull influences the inflow differently at different wind angles, and the performance can be both higher and lower at other wind angles. Moreover, the performance at for example quarterly wind is in many cases what makes a large difference between different wind propulsion technologies on the market. For those reasons, it is highly important to evaluate the wind propulsion technologies at other wind angles than 90 degrees.

By using the speed log, there is no need for double runs for the current correction. However, the double runs approach has the advantage that the disturbance of the hull to the wind anemometer can be checked and corrected for using the wind average method prescribed in the ISO/ITTC standards. Moreover, with double runs, the trial area can be kept to a smaller geographical area which means that wind, waves and water depth is more likely to be similar between the runs.

In the trial test presented in the present work, the rotor was alternative turn on and off for every double run. The gain of the rotor was derived by comparing two single runs with equal heading, with and without rotor. This approach will fail if the weather change during the trial so that the two runs are not comparable. For that reason, it could be a feasible approach to turn the wind propulsion on and off during each single run. This requires a larger undisturbed trial area. The approach will be tested by the authors in coming trials.

10. CONCLUSION

Wind assistance technology is one way to reduce the fuel consumption and green-house gas emissions from shipping. A key barrier for market uptake of wind assistance technology is the lack of verifiable data on the fuel savings potential [3]. So far there is no agreed standard for full scale verification of wind assistance technology of commercial vessels. In this work, a methodology based on short, dedicated trial is proposed. The trial is complemented with a route simulation and weather statistics to estimate the yearly fuel saving.

The methodology is demonstrated for the RoPAX ferry Copenhagen equipped with a Flettner rotor. A speed trial was conducted on the waters off Gedser in March 2021. The standard ISO/ITTC speed trial procedures were followed to as large extent as possible. In contrast to the normal procedures, the speed was measured using the ship's log and therefore no current correction was needed. The effect of the rotor was extracted by comparing single runs with and without rotor for the same wind condition.

Two methods to normalise the speed trial results are proposed. The first method uses the shape of the ships's speed power curve to extrapolate to nominal condition. This method involves several simplifications including the effect on propulsive efficiency due to changed propeller load. The second method is more complex and makes use of a ship simulation model. The difference between the results of the two methods are well within the estimated uncertainty margin.

It is demonstrated how a route simulation tool using hind-cast weather statistic can be used to extrapolate the trial result to yearly fuel saving for a specific route. The result for the specific case show that an average power saving of 4 % is possible. Continuous logging of wind speed on the route, the ship's fuel consumption and operability of the rotor for at least one year will complement this number and will be presented in future publications.

The mayor uncertainties include the disturbance of hull to the wind measurement onboard the vessel. This may disturb the relation between the trial result, which is based on the on-board measurements, and the route analysis that scale up the result to yearly fuel savings, which is based on the natural undisturbed wind on the ocean. Furthermore, the wind statistics introduce large uncertainties in the process. The largest uncertainty is probably the way the wind assistance technology will be handled and operated in reality. If the device will be in-active due to maintenance, failure, safety or other issues, then the power saving will off course be less. The same applies if crew choose to use the additional thrust from the wind to increase the ship's speed instead of reducing the power.

The proposed methodology is shown to be a feasible way to perform full scale verification for commercial vessels. With this approach a trustworthy result can be derived at a feasible cost, within a limited time frame, and using transparent, commercially available tools and established procedures.

The process will be refined further and applied to four other ships with wind assistance technologies. Among the issues that will be investigated is to keep the speed fixed over the runs instead of fixing the power, and to turn off and on the wind propulsion device within each run instead of between double runs.

11. ACKNOWLEDGEMENTS

The work is funded by EU Interreg North Sea Region.

12. REFERENCES

[1] International Maritime Organization, "Strategic plan for the organization for the six-year period 2018 to 2023," IMO International Maritime Organization, 2017.

[2] Rehmatulla, N., Calleya, J., Smith, T., "The implementation of technical energy efficiency and CO2 emission reduction measures in shipping", *Ocean Engineering*, Volume 139, pp 184-197, 2017

[3] MEPC.1/Circ.815 - 2013 Guidance on treatment of innovative energy efficiency technologies for calculation and verification of the attained EED (MEPC.1/Circ.815 (IMO, 2013)

[4] Paakkari, V., Hurford,. A., Craddock, C., "Rotor sail GHG reduction potential, modelling and sea trial validation" *Proceedings of 5th International Conference on Innovation in High Performance*

[5] *Sailing Yachts and Sail-Assisted Ship Propulsion,* Gothenburg, Sweden, June 2020

[6] Comer, B., Chen,.C., Stolz,.D., Rutherford, D., "Rotors and bubbles: Route-based assessment f innovative technologies to reduce ship fuel consumption and emissions" Working paper 2019-11, *The international council on lean transportation,* 2019

[7] Vahs, M., "Retrofitting of flettner rotors – results from sea trials of the general cargo ship "Fehn Pollux" *in Proceedings of the International Conference on Wind Propulsion*, 15th -16th October 2019, London, UK

[8] ISO 15016:2015 (2015), *Ships and marine technology – Guidelines for the assessment of speed and power performance by analysis of speed trial data*

[9] ITTC (2017b), Recommended Procedure 7.5-04-01-01, *Preparation, Conduct and Analysis of Speed/Power Trials*

[10] Øksnes, Ø., Dalheim, Steen, S., "Uncertainty in the real-time estimation of ship speed through water", *Ocean Engineering,* Volume 235, 2021

[11] Specialist Committee on Performance of Ships in Service, *Final report and recommendation to the 27th ITTC*, 27th International Towing Tank Conference 2014

[12] Strasser, G., Takagi, K., Werner, S., Hollenbach, U., "A verification of the ITTC/ISO speed/power trials analysis", Journal of Marine Science and Technology, 2015

[13] Kramer et.al, "Drift Forces –Wingsails vs Flettner Rotors", *High-Performance Marine Vehicles*, 2016

[14] Li, D.-Q., Leer-Andersen, M., & Allenström, B., "Performance and vortex formation of Flettner Rotors at high Reynolds numbers". *29th Synopsium on Naval Hydrodynamics.* Gothenburg. 2012

[15] Garenaux, M., Schot, J., Eggers,. R., "Numerical Analysis of Flettner Rotors Performances on the MARIN Hybrid Transition Coaster" *in the Proceedings of 12th Symposium on High-Performance Marine Vehicles, 2020*

[16] Jones, L., Prince, M., Hudson, D., Cocks, J., Predicted Fuel-Savings For A Flettner Rotor Assisted Tanker Using Computational Fluid Dynamics, In the Proceedings of RINA Wind Propulsion, 15th -16th October 2019, London, UK

[17] Norrbin, N. H. "Theory and Observations on the Use of a Mathematical Model for Ship Manoeuvring in Deep and Confined Waters", *Proc Eighth Symposium on Naval Hydrodynamics*, Pasadena 1970

[18] https://globalwindatlas.info/

[19] Werner, S., "Uncertainty of Speed Trials", in the *Proceedings of HullPIC*, 2020

[20] Insel, M., "Uncertainty in the analysis of speed and powering trials", *Ocean Engineering* 35, 2008

9. AUTHORS BIOGRAPHY

Sofia Werner is responsible for method development within fluid dynamics at SSPA. Her previous experience includes towing tank testing and ship design using CFD. 2011-2017 she was a member of the ITTC committee that developed the Recommended Procedures for Speed Trial. She is currently Chairman of the ITTC Specialist Committee for Wind Powered Ships.

Axel Hörteborn holds the current position of Specialist at SSPA Sweden AB. He is responsible for ship simulations, AIS data and Monte Carlo simulations analysis. His previous experience includes risk assessments of large-scale Monte Carlo inspired ship simulations, simulation of sailing ship.

Jonny Nisbet holds the current position of Senior Specialist at SSPA Sweden AB. His previous experience includes manoeuvring and seakeeping model testing, as well as simulation and CFD.

Rasmus Nielsen, Naval Architect in Scandlines. Educated Master Mariner and Naval Architect. Has been working 3 years as nautical officer, and one year at a consulting company before joining Scandlines in 2003. Working with projects in Scandlines, business case development and project management.

DEVELOPMENT OF A 1:30 SCALE SAILING MODEL OF OCEANBIRD

A Hillenbrand, **U Dhomé**, **J Kuttenkeuler**, KTH Royal Institute of Technology, Sweden
M Razola, Wallenius Marine AB, Sweden

SUMMARY

With the current developments of wind powered and wind assisted vessels comes the need for realistic ship model testing. Traditional test methods of free sailing models are somewhat limited by the difficulties to mimic the effects of the wind propulsion systems or at least for recreating them at the correct scale. Simulations are possible but time consuming and they cannot grasp all aspects that free sailing models comprise. At the KTH Maritime Robotics Laboratory, a 1:30 scale model of *Oceanbird* was developed and equipped with numerous sensors to enable accurate and realistic measurements while sailing in unrestrained waters. The paper presents the model's equipment and experimental results showing the abilities of the sailing demonstrator to gather high quality data for manoeuvring and aerodynamic studies. In particular, the definition of a zigzag manoeuvre for sailing ships is explained and results of tests are evaluated regarding manoeuvrability and aerodynamics.

NOMENCLATURE AND DEFINITIONS

AoA	Angle of attack ([-π π] rad)
AW	Apparent wind (-)
awa	Apparent wind angle ([-π π] rad)
c_{Fi}	Force coefficient (-)
cog	Course over ground ([0 2π] rad)
Fr	Froude number
p	Pressure (Pa)
Re	Reynolds number
SOG	Speed over ground (-)
TH	True heading ([0 2π] rad)
TW	True wind (-)
twa	True wind angle ([-π π] rad)
WPS	Wind Propulsion System
δ_R	Rudder angle (deg)
δ_{Wx}	Trim angle for wing x ([-π π] rad)
Δc_p	Differential pressure coefficient (-)
λ	Leeway ([-π π] rad)
ρ	Density [kg/m^3]

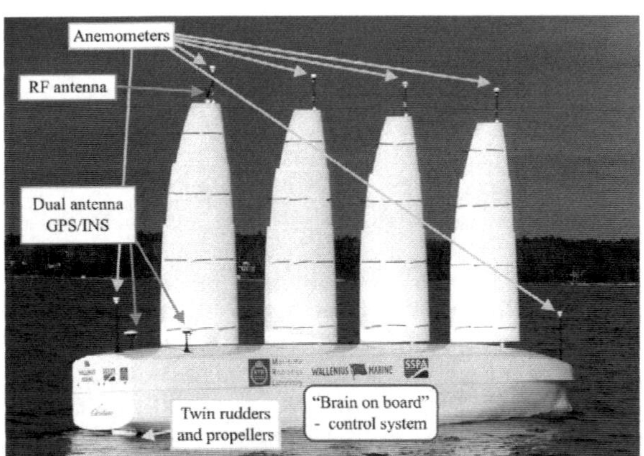

Figure 1: The 7-meter model sailing in Stockholm's archipelago. The third rig from the aft was for some tests exchanged with one that is fitted with pressure sensors, strain gauges and an extra anemometer mounted off the leading edge (see Fig. 2).

1. INTRODUCTION

Innovations towards sustainable shipping lead back to the most traditional way of propulsion on the oceans – sailing with wind propulsion. The Swedish concept *Oceanbird* developed by the main partners Wallenius Marine, KTH Royal Institute of Technology and SSPA Sweden AB involves rigid wingsails with symmetric profiles placed on deck along the centreline of a ~200 m long Ro-Ro ship capable of carrying about 7000 cars. In the research part of the project, several methods are used to predict the ship's performance in terms of aerodynamics [1], hydrodynamics [2] and seakeeping and manoeuvring [3,4]. For the latter, model tests in the towing tank and manoeuvring basin (MDL) of SSPA in Gothenburg were performed [3,4]. With the rise of wind propulsion for ships, the traditional towing tank and wave basin tests of self-propelled models need to be expanded to account for the aerodynamic forces generated by the wind propulsion system (WPS). This is a real challenge as hydrodynamic and aerodynamic forces scale very differently and the effects of the wind propulsion system are manifold: the aerodynamic forces are applied high up above the deck, meaning that not only thrust but also a large side force appears (generating drift and heeling moment) and also a yaw moment is generated, that can vary with the point of sail and trimming of the WPS. One way to tackle the problems in traditional test setups is to create a "wind tunnel" around a free-sailing model in a manoeuvring basin as tested in [5]. With such a setup, it is very difficult to obtain a good flow quality and the reachable Reynolds number is small, thus limiting the scalability. Another option is to mimic the effects of the WPS without representing the physics of it. One example is the towing of the model by winches placed at a specified height on a mast [6]. Another way, as presented in [7], is to replace the WPS by a fan or airscrew mounted on the model. These two methods only can replicate well the thrust and side force but rely on pre-calculated aerodynamic forces and neglect some of the effects of the WPS such as damping. For the experiments in the MDL at SSPA, an approach similar to [6] was chosen but with two individually controlled fans [3], which adds the ability to control the yaw

moment. The use of CFD enables to reduce some of the effects of the different scaling factors for better comparison with full scale results [4].
An alternative to model tests for seakeeping is full scale measurements in an uncontrolled environment. A real ship is equipped with an extensive set of sensors to measure the loads, motions, and environmental conditions, for example the "Berlin sail-force dynamometer" [8] from the TU-Berlin, the "Sailing Yacht Laboratory" [9] from Politecnico di Milano, the fitted J80 from IRENAV [10] or the measurements on the IACC yacht USA87 [11]. This type of experimental platform allows the use of a "normal" boat, but the accurate measurement of the environmental conditions makes the interpretation of the results somewhat difficult. The scale of the mentioned boats is appropriate for sailing yacht design – leisure or racing – but not for a merchant ship of much larger size as the *Oceanbird* project.

In an effort to fill the gap between model tests in a controlled environment and full-scale testing at sea, a fully sailing 1:30 scale model of the *Oceanbird* was built at KTH Centre for Naval Architecture. Previous projects in the laboratory proved the feasibility to accurately assess the performance and sailing behaviour of small sailing boats with simple electronic systems and mostly off-the-shelf sensors [12]. The 7 m long *Oceanbird* model is mainly used for the experimental assessment of aerodynamics, manoeuvring performance and control systems development. A sailing ship as tall and large as *Oceanbird* has never been seen before, hence the model also serves as a demonstrator for the general public to get acquainted with the project.

Table 1: Main particulars of the model

Length overall	7	m
Beam	1.3	m
Draught	0.284	m
Displacement	1165	kg
Wing height	2.67	m
Mean wing chord	0.77	m
Total wing area	8.18	m^2

The 7-meter model vessel (see Fig. 1) is equipped with a network of microcontrollers connecting environmental sensors and actuators. The closed control loop on board enables autonomous sailing, real-time monitoring of the onboard data, manipulation of control variables and data logging. The sensors are split in two categories: the ones making autonomous sailing possible (wind angles, heading, speed, among others) and a set of payload sensors, that are used for the aerodynamic evaluation of the boat and will later be used for advanced rig control. The payload sensors include pressure sensors on the wing skin and strain gauges on the mast.
The model hull is 1:30 scaled from the current design at the time of manufacturing. The design of *Oceanbird* evolved and is now fitted with 5 rigs instead of 4. This difference, however, does not affect the outcome of the research since the results are not meant to be scaled up directly. As discussed later in the paper, the scaling effects don't allow for that, instead the demonstrator is used to develop algorithms and sensors that can be used regardless of the size.

This paper first presents the hardware and the electronics architecture of the demonstrator. In the second part, some of the design challenges and a discussion around the scaling factors are issued. Finally, results from different tests are presented in order to give an overview of the possibilities and limitations of the demonstrator.

2. SYSTEMS OVERVIEW

2.1 MODEL

The model hull is made of a PVC foam covered with glass fibre and gelcoat. The four symmetric wings and the two rudders were manufactured in the same geometric scale as the hull at KTH. The wings are built as an assembly of an aluminium mast, extruded polystyrene foam ribs and glass fibre skin with vinylester resin. The rudders are made of a polystyrene foam core around a stainless-steel shaft with glass fibre and epoxy coating. The main particulars of the model are given in Table 1. A modular foundation frame in the bottom of the hull transfers the wing loads into the hull structure and accommodates the actuation mechanisms and electronics. The mass of the entire model when ready to sail is determined by the draught of the full-scale ship in loaded condition thus the draught is linearly scaled. In order to meet the stability requirements, the additional ballast is placed as steel plates at the lowest point inside the hull.

To enable semi-autonomous sailing, the boat is equipped with a variety of actuators and sensors, all integrated into a common on-board control system.

2.2 OPERATION

2.2 (a) Actuation

360 degrees of rotational freedom for each rig is accomplished with a stepper motor (Nema 34HS59-5004D) acting via a chain drive on the mast foot. Magnetic rotary encoders (Orbis BRD01_07) give feedback about the rig angle. The rudders are actuated jointly with a servo motor (Hitec HS-1100WP) through a pushrod-servo saver assembly. Like the full-scale ship, the model can be driven conventionally with two counterrotating two-bladed propellers. Two 24 V DC motors (MY1020) are coupled inside the hull to the propeller shafts.

2.2 (b) Sensing

Wind speed and angle are measured by ultrasonic anemometers (LCJ Capteurs CV7-E-OEM) at different locations on wings and hull. Anemometers are placed on each rig in extension to the pivot axis with a distance of 0.3 m to the wing tip. Two more anemometers are fixed to the hull above bow and aft. A dual antenna GNSS and AHRS (SBG Systems Ellipse2-D) is installed delivering position, true heading, course over ground and the boat motions.

2.2 (c) On-board computers

A distributed network of microcontrollers (tailored 32 bit 180 MHz ARM-Cortex-M4 processors based on Teensy 3.6) forms the "brain on board" consisting of a master-slave system with the superordinate "captain" and intelligent sensor bridges (ISB). Decisions *i.e.* target values, are taken only on the captain, while the ISBs act as a communication link to all actuation and sensing equipment and are placed in vicinity. Additionally, a Raspberry Pi 4 is integrated to the system to make use of the ROS framework and allow for applications that require higher computational power.

2.2 (d) Communication and Data Handling

A telemetry link on 800 Mhz (RF) enables real-time data monitoring and interaction with the on-board control system from a tender boat. Via a Wi-Fi link to the Raspberry Pi, the data can be visualized at a higher update rate. For manoeuvring in harbours, a radio control system is implemented, thus the boat can be operated with an RC controller.
Data is logged in two different ways for redundancy in case of failure. All relevant variables are written with a frequency of 10Hz on a SD card and on new data through a ROS framework.

2.2 (e) Operation

The implemented on-board control system allows for different modes of operation that can be manipulated via RF and RC from a tender boat. According to the chosen "task", the boat sails on a desired course or with a constant apparent wind angle. Currently, sail trimming and steering are controlled independently from each other.
Steering: The rudder angle is obtained from a PID controller with the course error as input. When sailing on a desired course, the course error is calculated as the difference between measured course and target course. The control parameter "course" can be chosen via RF either to be the course over ground (*cog*) or the true heading (*TH*). For sailing to a constant apparent wind angle (*awa*), the course error is simply the difference between desired and measured apparent wind angle.
Sail trimming: The trim angles of the sails are set such that the wing experiences an inflow under a desired angle of attack (*AoA*). A forward force is generated when *AoA* is positive for positive *awa*, thus when sailing on a starboard tack. The desired angle of attack can be manipulated via RF. The crucial control parameter for sail trimming is the apparent wind angle. It is measured at different locations on the boat, therefore several options of using the sensed wind can be chosen. Individual trim angles for the four wings are obtained when considering each rig with its own wing-fixed anemometer as an individual unit. In contrast, choosing *awa* from a single anemometer (*e.g.* the windward one) or an averaged *awa* from several anemometer readings as common trim control value results in the same trim angle for all wings. When desired, the wings can be depowered by trimming for an angle of attack of zero resulting in a zero-lift force for symmetric wings.

2.3 PAYLOAD SENSORS

In addition to the sensors and actuators necessary for the operation of the model that are described in Section 2.2, a test rig was especially instrumented to further investigate the aerodynamics with pressure sensors and strain gauges, as depicted in Figure 2.

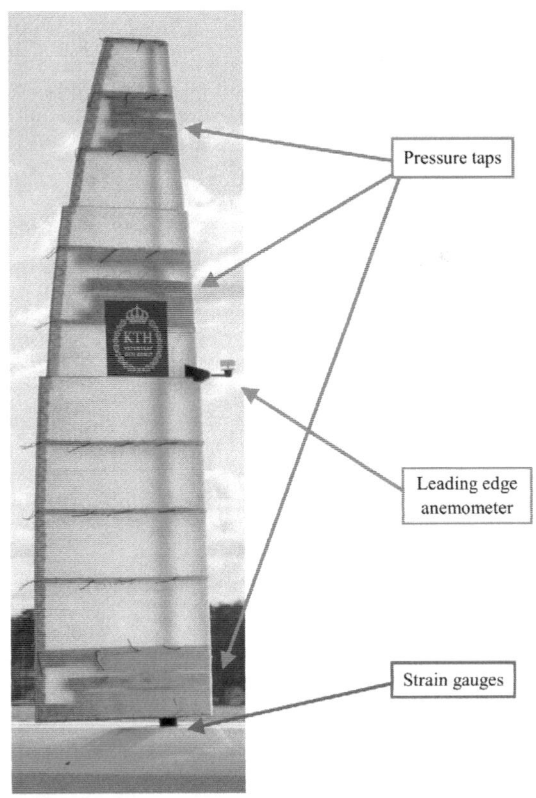

Figure 2: The extra instrumented wing, showing the location of the pressure taps, the strain gauges and an anemometer.

2.3 (a) Pressure Sensors

The pressure difference Δp between the suction and pressure side of the wing is directly measured with MEMS based differential pressure sensors (Honeywell HSCDRRD002NDSA3). The sensors are spread along three strips placed at 5%, 62.5% and 87.5% of the wingspan measured from the root. The lower two strips contain seven sensors with the taps placed at 3%, 10%, 18%, 31%, 44%, 57% and 70% of the chord from the leading edge. The uppermost strip contains 6 sensors with taps at 3%, 10%, 18%, 35%, 52% and 70% of the chord.

The sensors are capable of measuring pressure at around 2 kHz. Each strip of sensors is controlled by one Teensy 4.1 board, which reads the sensors consecutively at a rate much higher than the update rate of the sensors, so that all sensors can be considered as read as soon as a new value is available. 200 samples are averaged, and the values are sent to an ISB controlling the sensors in the whole wing. Thus, the pressures are effectively measured at 10 Hz.

2.3 (b) Strain Gauges

A total of six strain gauges realizing three full Wheatstone bridges are placed just in-between the upper bearing at the deck level and the root of the wing. The strain measurements are therefore not influenced by the lower mast construction and wing actuation inside the boat. The linear deformations are measured in the wing chordwise (x_W, see definitions in Fig. 3) and transverse direction (y_W), while the third bridge measures the torsion on the mast. The signal from the bridges is amplified with HX711 load cell amplifiers from Sparkfun.

In similar manner as for the pressure sensors, all amplifiers are controlled by one Teensy 3.6 with data acquired from the gauges at 80 Hz. An average of 8 samples is sent to the ISB to reach an effective frequency of 10 Hz.

Calibration performed in the lab allows the conversion from deformation to actual moments. An assumption needs to be made on the spanwise location of the centre of effort to convert the bending moments into forces. For the torsion of the wing, since the mast is located at 25% of the chord, nothing should be measured as long as the effective angle of attack on the wing is small (*i.e.* in the linear region of the lift curve). Stall, or at least exiting of the linear region can then easily be detected.

3. SCALING AND STABILITY

3.1 HYDRODYNAMIC AND AERODYNAMIC SIMILARITY

Performing tests with a geometrically scaled sailing model outside a laboratory without a controllable environment poses various challenges. Ideally, model (*M*) and full-scale prototype (*P*) should be similar in geometry, kinematics and dynamics such that the model test results involve no scale effects. As described in Section 2, the demonstrator is geometrically similar to the full-scale vessel with a scale factor of 30. For a sailing ship, relevant force ratios are the Froude number *Fr*, relating inertia and gravity, and the Reynolds number *Re,* which is the ratio of inertial forces to viscous forces. For conventional ships, the Reynolds number is the one affecting hull and appendages, mostly considered for drag. In the case of a sailing ship however, the aerodynamic Reynolds number on the sails and hull is of primary importance. Often sailing model tests are therefore performed both under Froude similarity $Fr_M = Fr_P$ in towing tanks to study the hydrodynamics where gravity and inertial forces are dominant and in wind tunnels under Reynolds similarity $Re_M = Re_P$ to study the aerodynamics, where viscous forces are of great importance. As the presented model is tested in an uncontrollable environment, both force ratios cannot be kept identical to the full-scale ratio at the same time. However, by limiting the tests to relatively low wind speeds, Froude similarity can be achieved. For the aerodynamic part, since the demonstrator evolves in the same air and at similar wind speeds, the chord of the wings would need to be almost the same as at full scale to obtain the correct aerodynamic Reynolds number. Similarly for the rudders, the chord would need to be larger than the one of the full-scale ship for the hydrodynamic Reynolds number to match, since the model has a lower speed than the full scale ship to maintain the Froude ratio.

Table 2 shows a comparison of the different ratios and characteristics at full scale and model scale. Note that different speeds are used in the calculation of the aerodynamic Reynolds number. In full scale, a mean apparent wind speed of 10 m/s is used throughout the design process. For the model scale, however, a lower value of 6 m/s is used that represents in average the apparent wind speed needed to propel the boat at model boat speed from Froude similarity.

Table 2: Main characteristics at full scale and model scale

	Unit	full scale	Model scale
Boat Speed	[kn] [m/s]	10 5.14	1.87 0.96
Froude Number	[-]	0.116	0.116
Rudder Reynolds number	[-]	~3.8 *10^7	~4.7 *10^5
Wind speed	[kn] [m/s]	19.4 10	11.6 7
Aerodynamic Reynolds number	[-]	~15 *10^6	~30 *10^3

As shown in Table 2, some of the ratios are very different, which implies potentially large differences in behaviour of the model and the full-scale ship. However, most effects arising from these differences can be

predicted. The main effect of the lower Reynolds number is that the maximum lift coefficient is expected to be lower and stall angles should be lower. In the linear region of the lift curve, no major difference should be present. For the rudders, this means that for a similar deflection, the model scale will likely generate smaller side force than the directly scaled force. The deflection at which the rudder will stall is also expected to be smaller than at full scale. These effects have been predicted using CFD by SSPA in the scope of the *Oceanbird* project [4].
The interaction effects on the wings are not yet fully understood, and a better comprehension is part of the *Oceanbird* project [1]. However, the main expected effect of the Reynolds number on the interaction is that each wing individually will see its stall angle reduced at model scale because of the lower Reynolds number but the wing-to-wing interaction should remain similar, considering that the current studies show that it mostly depends on inertial effects rather than the viscous ones. The tip vortices at the end of the wings are also likely to be modified by the Reynolds effects, but their prediction is very complex. Traditionally in towing tanks or wind tunnels, the effects of viscosity can be mimicked even at low Reynolds number by forcing the boundary layer on the hull or wing to be turbulent by tripping it with extra roughness. This is difficult to realise in a non-controlled environment because precise design of the tripping system depends on the fluid speed. It is also difficult to predict whether the effect of tripping the boundary layer is exactly similar to it naturally becoming turbulent under high Reynolds number. It was thus decided not to try to trip the boundary layers in any specific ways. The rudders are handmade and have a quite rough surface, so the boundary layer is likely not as laminar as it should be at this Reynolds number.
From this we can conclude that for the aerodynamic side, the effects of lower Reynolds number will prevent direct comparison with full scale predictions, but a comparison of different wing trimming strategies on the demonstrator should be valid. For the hydrodynamic side and ship manoeuvrability, a similar conclusion can be drawn: the rudder angles and leeway angles will differ from full scale, but comparison of different manoeuvres or trimming strategies are possible.

3.2 STATIC STABILITY

Sea trials with the model are performed in the Stockholm archipelago (more precisely Baggensfjärden and Stora Värtan) in the Baltic Sea. The wind and sea conditions do not necessarily match a scaled version of what the prototype will encounter on the oceans and can only be 'controlled' to the extent that tests are scheduled such that they can be performed in suitable conditions.

The sailing model is both a research platform for sailing physics and performance and a proof of concept that shows this novel type of cargo vessel when sailing. In contrast to traditional sailboats of the same length, large static heel angles are for the model not appropriate in order to mimic the sailing appearance of the prototype. During the development process, requirements have thus been specified to match the expected environmental conditions to be encountered:

- The demonstrator sails in conditions of *0-6 m/s* true wind speed with a static heel not exceeding 10 degrees.
- The model is designed to survive *12 m/s* true wind speed.

The requirements guided the transverse stability considerations. Since draught and displacement are fixed as given in Table 1, the static heel and thus transverse stability in terms of lever arm *GZ* are adapted with the vertical centre of gravity.

3.3 WAVES

One important environmental aspect, that cannot be controlled, is the sea state. Because of the small scale of the model, even small waves of just a dozen centimetres height can induce large errors and represent a high sea state. Fortunately, in the relatively closed bays where the tests take place, the waves are usually very small if not non-existent in the wind conditions mentioned before and their influence was thus not evaluated further at this stage.

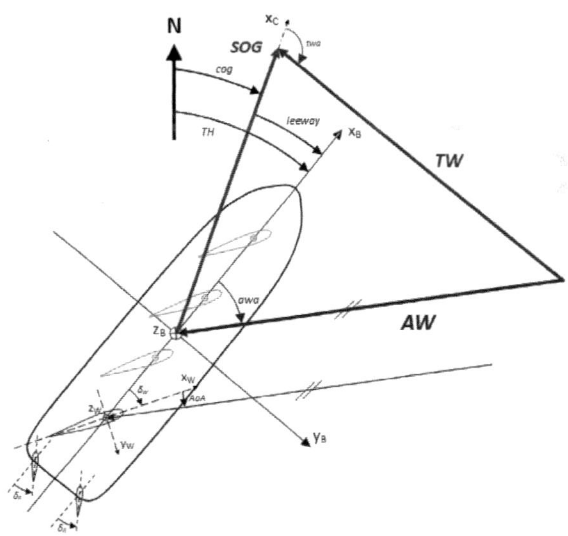

Figure 3: Coordinate systems and definitions

4. TEST RESULTS AND DISCUSSION

This section presents test results from sailing that took place in spring 2021 dedicated to manoeuvring performance. As an example, Fig. 4 shows the trajectory of a test day. Earlier tests were performed to tune control parameters enabling semi-autonomous sailing and more advanced manoeuvres. Additional tests were performed to test and collect data with the extra instrumented rig (see explanation in Section 2.3). For the latter, the third rig from the aft was replaced by the instrumented one as shown in Fig. 2. On a very calm day, extra manoeuvres were performed without the wings, using only the motors for propulsion.

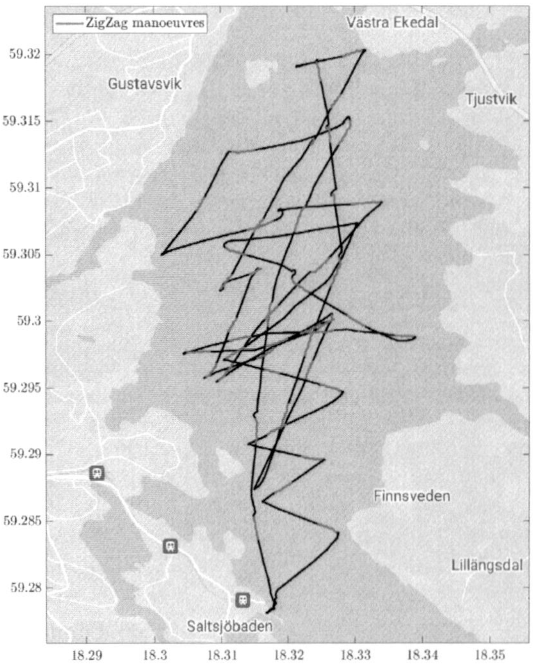

Figure 4: Example of the trajectory during a test day in Baggensfjärden, in Stockholm's archipelago. This specific day was dedicated to zig-zag manoeuvres (shown in red on the track).

4.1 MANOEUVRING

An important use of the demonstrator is to evaluate the manoeuvring performance of such sailing ships and in particular to evaluate and quantify the effects of sailing on manoeuvrability. To date, ships must comply with the IMO regulation regarding manoeuvrability, but it is still unclear if the standard manoeuvres (zigzags, turning circles, etc) should be executed for wind powered and wind assisted vessels while sailing or on motor power and if the requirements remain the same as for vessels without wind propulsion. In the ABS "Guide for wind assisted propulsion system installation" [13], it is stated that "The manoeuvrability of the vessel is to be verified at sea trials (...)", but there is no mention on how the tests should be performed. The IMO requirements give a specific regime for the engine to perform the tests at; for pure sailing manoeuvrability, this is not applicable. Due to the fluctuating nature of the wind, one can also expect the speed of a sailing ship to be varying, which limits the possibility to set a requirement on approach speed during standard manoeuvres. Another question that arises is how the control systems of the WPS should be used during manoeuvres. In the case of *Oceanbird* and the 7 m model, the rotation of the wings for example, might not be as fast as the rotation of the ship during a zigzag; should the wings remain fixed, should they move anyway? This is yet to be decided.

Tests to measure the influence of such parameters were performed with the 7-meter model. Specific definitions and choices had to be adopted, as well as a limited choice of standard manoeuvres. As a first attempt, the focus was on zigzag manoeuvres to assess the yaw-checking ability. Both 10/10 and 20/20 zigzags were executed, with different control strategies. The reader is referred to [14] for the definition of the zigzag manoeuvre.
Since scaling to full scale is not directly possible (see consideration in Section 3.1), a reference case needs to be established to compare the sailing manoeuvres with. This was done by performing manoeuvres without the wings, using only the electric propulsion. These tests were performed at different speeds and in both port and starboard directions during a day with very low wind.

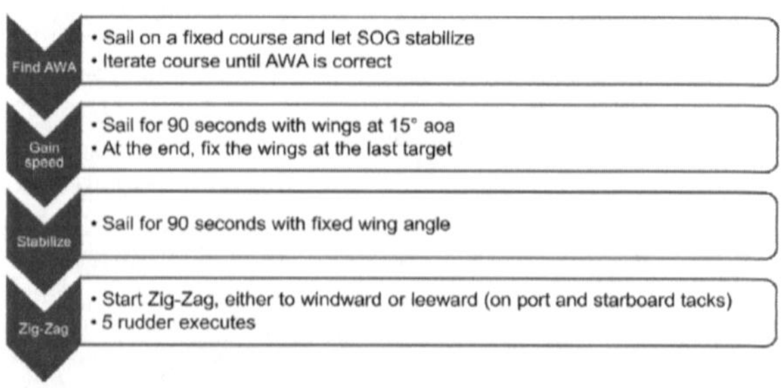

Figure 5: Flowchart explaining how the zigzags manoeuvres were performed.

The true wind angle (*twa*, see Fig. 3 for the definitions) has a large influence on the behaviour of any sailing boat, therefore it has likely a large effect on manoeuvrability. However, from an aerodynamic point of view, it is the apparent wind angle (*awa*, see Fig. 3) that has the largest influence. All the zigzag manoeuvres were thus performed at three different starting *awa* (45°, 90°, 135°) rather than *twa*.

As a first attempt to evaluate the impact of sailing on manoeuvrability, it was chosen to keep the number of varying parameters as low as possible. At the beginning the tests were therefore performed with the wings fixed with respect to the boat during the whole manoeuvre. With a known wing angle and varying boat heading, it is easier to interpret the variation of the aerodynamic forces. On the full-scale ship, it is reasonable to believe that the wings won't be able to rotate very fast, hence fixed wings will be closer to reality. The manoeuvre proceeds as follows, a schematic summary is shown in Figure 5.

- Preparation: The boat sails on a straight course until the speed is stabilized. The course is iterated until the *awa* matches the chosen target *awa*.

Once on an appropriate course, the standard manoeuvre is performed:

- Phase 1: The boat sails for 90 seconds on the chosen course with the wings adapting to the incoming wind (keeping an *AoA* of 15 degrees) to maximize thrust. The boat gains speed and the *SOG* stabilizes. In this phase, a moving average is calculated over the wing target angles.
- Phase 2: The wings are fixed at their average position calculated during phase 1 and the boat sails for another 90 seconds on the same course.
- Phase 3: The actual zigzag manoeuvre with five rudder executes is performed. After the last execute, the boat sails for another 30 seconds with the rudder fixed at its initial position.

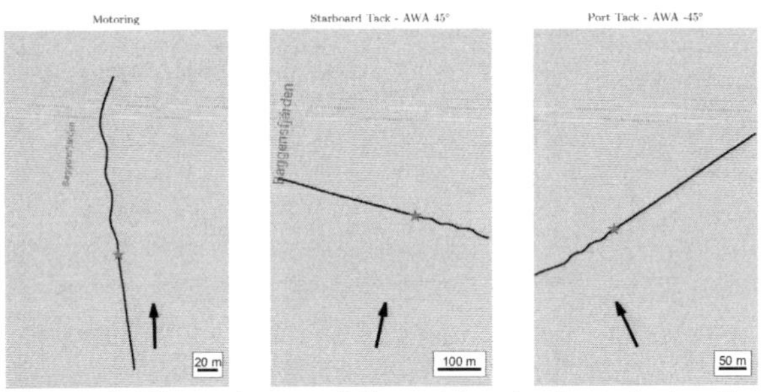

Figure 6: The trajectory followed by the boat during 3 different zigzag manoeuvres, executed at a similar approach speed. On the left, a motoring test. In the middle, the boat is sailing on starboard tack at awa ≈ 45°, on the right, it is sailing on port tack at awa ≈ -45°. The thick arrow on the bottom shows the average true wind direction during the time preceding the manoeuvre. The star marks the transition between phase 2 and 3, i.e. the actual start of the zigzag manoeuvre.

The results of three zigzag manoeuvres performed at a similar approach speed are shown in the Figures 6 to 8 with the trajectories depicted in Fig. 6, the motoring case results in Fig. 7 and two cases at awa ≈ ±45° in Fig. 8 (one on port tack and one on starboard).

From Fig. 8, when looking at the part before the zigzag start, one clear result is the large amount of leeway that the boat experiences when sailing upwind (here $\lambda \approx 15°$). This very large angle is likely overestimated at this scale compared to what the full-scale ship will experience due to scale effects of the lift and drag on the appendices [4], but a fairly large amount of leeway angle is to be expected for such sailing ships.

A very clear difference between the motoring and sailing cases is the behaviour of the *SOG* during the zigzags. When motoring, the curve looks as expected, with a slight decrease of speed at the beginning, followed by small oscillations around a rather constant value. In the sailing cases however, the speed is very different as it follows the variations of apparent wind angle very closely. Since the wings are fixed during the manoeuvre, this result is expected and easy to interpret: as the boat turns towards the wind, the angle of attack *AoA* is reduced thus lowering the thrust, as can be seen in the decrease of *SOG* between t = 0 s and t ≈ 15 s in Figure 8 on port tack (bottom graph). But when turning away from the wind (to leeward) the opposite effect happens as the increase of SOG proves e.g. during the first rudder execute on starboard tack in Fig. 8 (t = 0 to t ≈ 15 s in the second plot from top). If the wings were not kept at a constant angle, the speed is not expected to vary like that.

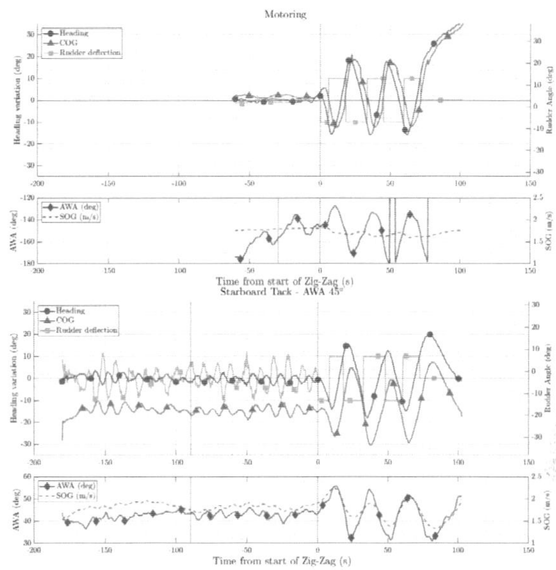

Figure 7: Result of a zigzag manoeuvre with the boat motoring. The first rudder execute is performed to port. In the upper part, the heading variation is shown (blue line with round marks) together with the variation of course over ground (red line with triangles) on the left axis. On the right axis, the rudder angle (yellow line with squares) is shown. The horizontal line shows the initial rudder value to maintain the heading. The vertical line at t = 0 s shows the beginning of the zigzag manoeuvre. In the lower part, the apparent wind angle is shown on the left axis (blue line with rhomboids) and the speed over ground (red dashed line) is shown on the right axis.

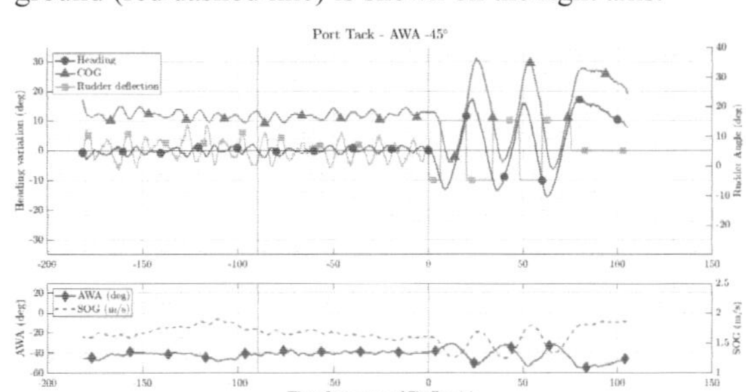

Figure 8: Results of zigzag manoeuvres when sailing upwind on starboard (upper graph) and port (lower graph) tacks, with the first rudder execute to port. The legend is similar to the one in Fig. 7. It should be noted that the scale of the rudder angle axis is the same on all graphs, but the zero value is not aligned with the heading variation, which shows the constant rudder deflection needed to sail on a straight course before the start of the zigzag.

In the sailing cases, even if the speed varies, its average is still close to the approach speed, which leads to a total duration of the manoeuvre remaining very close to the duration of the motoring case. One important difference, however, is that the duration of each rudder execute varies more in the sailing cases. This can be due to small differences in the wind, but also due to the difference in terms of aerodynamic forces from increasing or decreasing the angle of attack. The repeatability of the sailing manoeuvres is highly affected by these effects, and the cases shown here were specifically picked because the manoeuvres were performed smoothly.

It is worth noting that the cases presented here have a speed significantly larger than the target speed presented in Table 1, due to larger wind speeds. Here again, since the manoeuvres were performed at similar speeds, case-to-case comparison is possible.

4.2 PRESSURE AND LOADS

Some tests were performed with the wing that is instrumented with pressure sensors and strain gauges (s. Section 2.3). Initial results collected in zigzag manoeuvres are presented below. The data shown in Fig. 9 was recorded during a zigzag manoeuvre with fixed wings, thus as the boat heading changes, the angle of attack increases or decreases. *AoA* is plotted in yellow marked with stars in the graph on the bottom of Fig. 9, together with *TH* (blue line with dots), *awa* (purple line with triangles) and the rudder angle δ_R (orange line marked with squares).

Using the differential pressure recordings, a differential pressure coefficient Δc_p can be calculated as follows:

$$\Delta c_p = \frac{\Delta p}{0.5 \cdot \rho \cdot aws^2}$$

This value is calculated per pressure tap in all strips and plotted over time in the upper three graphs of Fig. 9.

The forces recorded with the strain gauges on the mast can be nondimensionalized in a similar way:

$$c_{F_i} = \frac{F_i}{0.5 \cdot \rho \cdot aws^2 \cdot S}$$

Note, F_x and F_y denote here the forces converted in the x-respective y-direction of the boat fixed coordinate system (in Fig. 3 marked with x_B and y_B). The strain gauges give a measure of the moments at the root of the mast, not of forces directly. To derive the forces, a fixed location of the centre of effort was assumed in the middle of the wing span. For the nondimensionalization, an air density of ρ = 1.225 kg/m^3 was assumed and the wind speed measured with the anemometer placed in front of the leading edge was used (in the equations denoted as *aws*).

The influence of the changing *AoA* can clearly be seen in both the plots of the differential pressure coefficients and the force coefficients (see Fig. 9). Maximum values of Δc_p are reached when the absolute value of *AoA* is smallest. Note, the boat is sailing here on a port tack, thus the *AoA* must be negative for the wings to generate lift propelling in forward direction. With the boat turning towards starboard (*TH* increases, positive rudder angles δ_R), the absolute value of *AoA* increases to values of up to 45 degrees and the wings stall, which is represented by a collapse of the differential pressure profiles. When turning towards port, the *AoA* reaches again a range where the flow reattaches and the Δc_p values closest to the leading edge increase again. An interesting effect that can be seen in Fig. 9 is the earlier stall near the tip of the wing, as can for example be seen between 25 and 50 seconds. On the upper part of the plot, the pressure curves are almost collapsed, while they are clearly not in the lower section of the wing.

The measurements of *AoA* should be considered carefully; indeed, as can be seen in Fig. 2, the anemometer is placed very close to the leading edge of the wing, at a location where the flow is likely influenced by the wing. This explains the very large values of *AoA* measured.

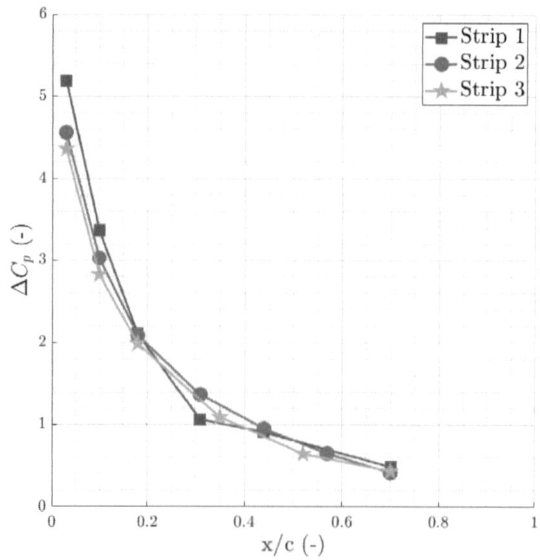

Figure 10: Differential pressure coefficient Δc_p profiles for strip 1 to 3 at the time marked with a vertical dotted black line in Fig. 9.

An example of Δc_p profiles is depicted in Fig. 10. The profiles for the three pressure strips (strip 1 is the lower most one) correspond to the time indicated with a black vertical dotted line in Fig. 9. As expected, the value of Δc_p is here largest close to the leading edge and decreases towards the trailing edge.

The force coefficients can seem a little bit counterintuitive. Indeed, one would expect most of the forces to be in the boat's longitudinal axis rather than the transversal axis while sailing. The case presented here corresponds to an upwind tack with *awa* varying around -50°. The very large leeway shown in Fig. 8 is the result of a very large side force. The time variation of both forces however are in good accordance with the variation of pressure.

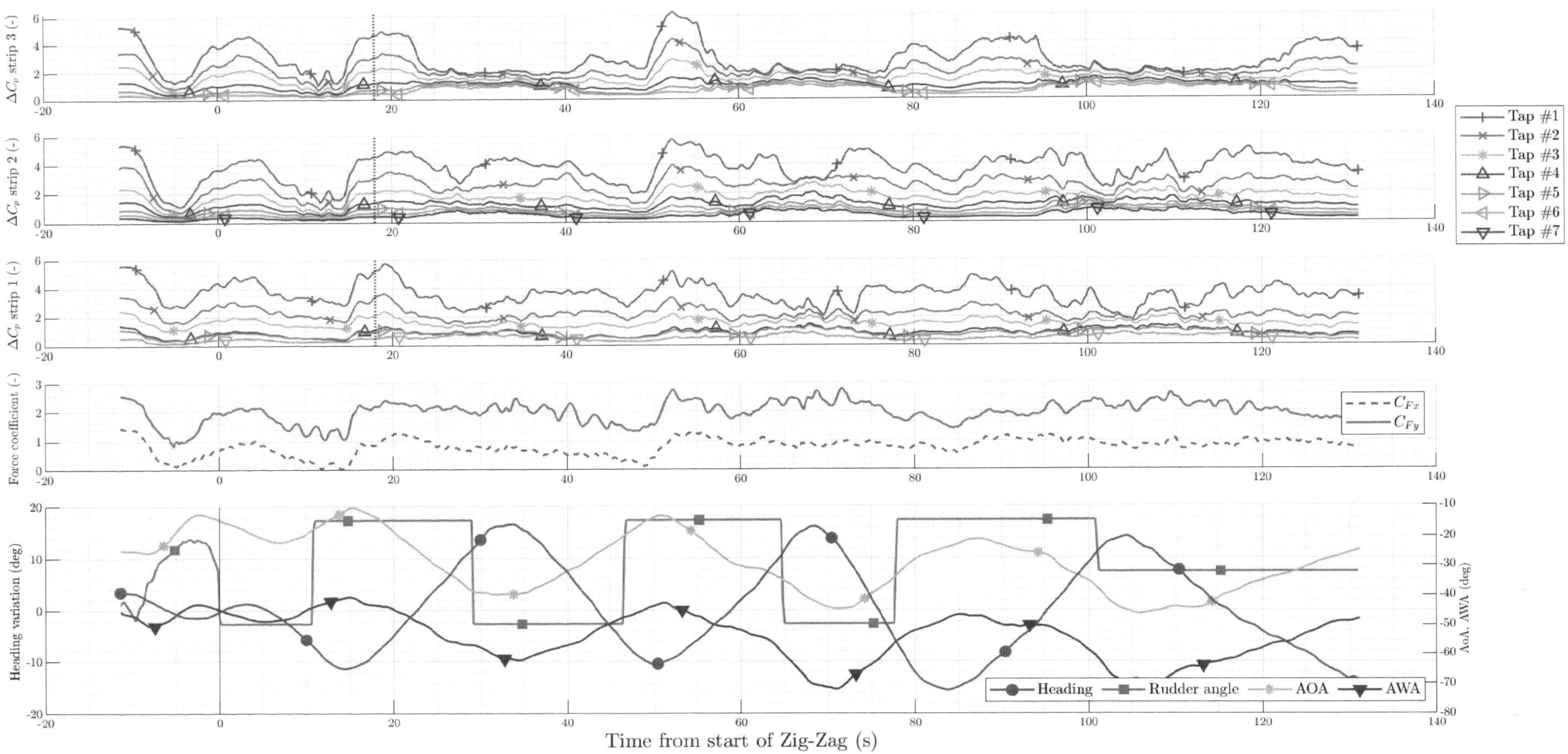

Figure 9: Payload sensor data plotted against time obtained in a zigzag manoeuvre with the extra instrumented rig (see Section 2.3) - differential pressure coefficient Δc_p of the three strips on the wing (upper three plots), force coefficients c_{Fx} and c_{Fy} in the boat fixed coordinate system (in Fig. 3 denoted as x_B and y_B) and major variables (*AoA*, *awa*, δ_R and *TH*, see definitions in Fig. 3) of the zigzag manoeuvre (bottom plot). This type of plot for the pressure coefficient against time is not commonly used, but here it was chosen because it gives a good idea of the temporality of the signals. The most common representation of the pressure is to look at the variation at fixed chordwise locations; this can be done by looking at "vertical slices" of the plot: the more space there is between the lines, the steeper the pressure profile is. The apparent wind speed is measured in front of the leading edge (see Fig. 2).

The presence of both pressure sensors and strain gauges, even if partially redundant, enables one to gain a clearer idea of the aerodynamic effects. For example, between 100 and 120 seconds, the tip of the wing is stalled, but almost no effect is visible on the forces; this shows that only a small area of the wing must be affected by the stall, thus limiting the overall effect. With an increased span-wise distribution of pressure sensors, the location of the aerodynamic centre of effort could be estimated, making the force estimates more accurate.

5. CONCLUSIONS

The paper presents the development of a free sailing 1:30 scale model of *Oceanbird* that is used for studies on manoeuvrability, aerodynamics and control systems development. Early experimental results presented in the paper prove the ability to accurately measure and acquire main parameters for boat motions and flow characterization to assess manoeuvrability and influencing aerodynamic effects. The scalability of the results from a model evolving in a non-controlled environment is discussed to present the limits and also the relevance of this approach.

The current regulations on ship manoeuvrability do not specify how standard manoeuvres are to be performed for wind powered vessels, hence the authors propose one way of performing zigzag manoeuvres. The results for these experiments show that the wind conditions as well as the control algorithms for the WPS have a large influence on the boat's manoeuvrability. The pressure and force sensors on an extra instrumented wing provide promising results and will be further investigated in regard to the possible use to improve the existing control algorithms.

6. ACKNOWLEDGEMENTS

This work is supported by the Swedish Transport Agency grant TRV 2018/96451.

7. REFERENCES

[1] MALMEK, K, et. al., 'Comparison of two rapid numerical methods for predicting the performance of multiple rigid wing-sails', Proceedings of 5th International Conference on Innovation in High Performance Sailing Yachts and Sail-Assisted Ship Propulsion, Gothenburg, Sweden, June 2020.

[2] MARIMON GIOVANNETTI, L., et al., 2020. 'The effects of hydrodynamic forces on maneuvrability coefficients for wind-assisted ships.' Proceedings of the ASME 2020 39th International Conference on Ocean, Offshore and Arctic Engineering, Fort Lauderdale, USA, July 2020.

[3] GERHARDT, F. & SANTEN, V., 2021. 'Seakeeping model testing for a wind powered vessel.' online, RINA Wind Propulsion Forum, London, UK, February 2021.

[4] MARIMON GIOVANNETTI, L., et. al., 'The art of model testing: Using CFD to adapt traditional tank testing techniques to a new era of wind propelled shipping', The 9th Conference on Computational Methods in Marine Engineering (Marine 2021), online, June 2021

[5] EGGERS, R. & KISJES, A. S., 'Seakeeping and manoeuvring for wind assisted ships'. RINA Wind Propulsion conference, London, UK, October 2019

[6] EGGERS, R., 'Prediction methods for Team Akzo Nobel in the Volvo Ocean Race and Wind Assisted Ship Propulsion for merchant vessels.' International HISWA Symposium on yacht design and yacht construction, Amsterdam, The Netherlands, November 2018.

[6] GAUVAIN, E., 'The Un-restrained Sailing Yacht Model Tests - A New Approach and Technology Appropriate to Modern Sailing Yacht Seakeeping.', The 23rd Chesapeake Sailing Yacht Symposium, Annapolis, USA, March 2019.

[8] HOCHKIRCH, K. & BRANDT, H., 'Fullscale Hydrodynamic Force Measurement on the Berlin "Sail–Force–Dynamometer".', the 14th Chesapeake Sailing Yacht Symposium, Annapolis, USA, January 1999

[9] FOSSATI, F., et. al., 'A novel full scale laboratory for yacht engineering research.', Ocean Engineering, Vol. 104, pp. 219-237, 2015

[10] AUGIER, B., et. al., 'Experimental validation of unsteady models for fluid structure interaction: Application to yacht sails and rigs.', J. Wind Eng. Ind. Aerodyn., Vol. 101, pp. 53-66, 2012

[11] BINNS J.R., et.al., 'The development and use of sailing simulation for IACC starting manoeuvre training', 3rd High Performance Yacht Design Conference, Auckland, New-Zealand, December 2008

[12] DHOME, U., et.al., 'Energy efficient self-steering mechanism for an autonomous sailing vessel.', OCEANS 2019 Marseille conference, Marseille, France, 2019

[13] American Bureau of Shipping, 'Guide for wind assisted propulsion system installation', June 2020.

[14] American Bureau of Shipping, 'Guide for vessel manoeuvrability', February 2017.

8. AUTHORS BIOGRAPHY

Antonia Hillenbrand is a master student focussing on fluid dynamics and naval architecture at KTH Royal Institute of Technology and TUM Technical University of Munich. She was previously working on wind tunnel testing and CFD and was involved in the development especially the control of the 7-meter model at KTH.

Ulysse Dhomé is a PhD student at the Centre for Naval Architecture at KTH Royal Institute of Technology, Stockholm. His previous experience includes experimental and numerical work on sail aerodynamics as well as development of sailing robots.

Jakob Kuttenkeuler holds the current position of full professor at the Centre for Naval Architecture at KTH where he e.g teaches and is responsible for the Maritime Robotics Laboratory. His previous relevant experience includes numerical/experimental research on aeroelastic optimization, robotic sailing, ship routing using ensemble weather forecasts, performance analysis of sailing vessels, parametric roll instabilities etc.

Mikael Razola is the technical director for the Oceanbird Sailtech project at Wallenius and project manager of the Wind Powered Car Carrier research project since 2019 working closely with SSPA Sweden AB and KTH Royal Institute of Technology on various research topics related to sailing ships.

THE HORNS OF THE TRILEMMA: SEAKEEPING MODEL TESTS FOR A WIND POWERED VESSEL

F C Gerhardt, M Kjellberg, I Wigren, S Werner, SSPA Sweden AB, Sweden
M Razola, Wallenius Marine, Sweden

SUMMARY

Assessing seakeeping performance at an early stage is even more important for wind powered vessels than for conventional ships, since there is little design experience to lean on. When the driving force comes from sails instead of a propeller, ship dynamics change considerably. Course keeping, turning ability, motions and acceleration in waves are just some of the properties that must be assessed. Including wind propulsion devices in a model test is however not straight forward. In this paper we present a methodology for model testing wind powered vessels. *Rpm* and azimuth-controlled fans/airscrews are used to mimic the aerodynamic forces from the sails. Results from model tests with a car carrier are presented and discussed while particular attention is paid to possible improvements of the test methodology.

1. INTRODUCTION

In April 1926 a young British scientist called Hermann Glauert put the finishing touches to a textbook he called "The elements of aerofoil and airscrew theory" [1]. The book would firmly establish his reputation as one of the leading experts in the emerging field of "aerodynamics".

Less than two month later, in June 1926, a four-masted steel barque with the name *Padua* was launched at the Tecklenborg yard in Bremerhavn. She was to be the last of the famous flying P-liners run by the shipping company Laeisz out of Hamburg. After less than 20 voyages to Australia and Chile she was decommissioned at the outbreak of the second world war.

While the *Padua* eventually became a symbol for the bygone area of the windjammers the ideas presented by Glauert are still widely used today. Through his book the English-speaking world became acquainted with aerodynamic concepts developed by Ludwig Prandtl's research group in Germany. Having published his "boundary layer" theory in 1904 Prandtl had attracted a number of talented students to the University of Göttingen. Among them household names like Birnbaum, Ackermann, Betz, Kármán, Munk and Trefftz. A century later the two pillars of the Prandtl/Glauert wing design methodology, "Lifting Line" and "Thin Aerofoil Theory", still stand and contribute to our understanding of "lifting surfaces", why they generate lift, and how to minimise their drag.

Sailing ships, however, would benefit very little from these new insights. As the fate of the *Padua* shows, their days were numbered before engineers began to properly understand the physics of sailing. Out of necessity the ships that had ploughed the seas for thousands of years were built based on empirical knowledge and hard-won experience, not scientific principles.

1.1 BACKGROUND

Due to the many advantages of Diesel propulsion and the availability of cheap oil interest in commercial sailing vessels has been minimal for most of the 20th century. Now, at the beginning of the 21st century the climate crisis has changed this.

In 2018 international shipping accounted for around 2.89% of globally produced CO2 emissions. Forecasts about a future increase are grim: From 90% to 130% by 2050 compared to the baseline year of 2008. Such disquieting projections have placed the topic of decarbonising shipping on the forefront of the global policy making agenda. The International Maritime Organization (IMO) reached a milestone agreement in April 2018. In alignment with the emission reduction goals set out in United Nation's 2015 Paris Agreement, IMO agreed to cut greenhouse gas emissions by at least 50% by 2050 and CO2 emissions per transport work by at least 40% by 2030 and 70% by 2050 compared to the 2008 level [2].

To achieve such ambitious goals, the international community is currently examining measures and possible actions to enhance the energy efficiency of the maritime transport sector. Among the various options to reduce fuel consumption and CO2 emissions IMO has started to pay attention to the exploitation of wind assisted ship propulsion (WASP). Several WASP-concepts are currently being developed and re-discovered. Such modern technologies have little in common with the square sails of the *Padua* and range from "Flettner-rotors" over kites and suction sails to rigid sails that resemble vertical aircraft wings [2].

1.2 WIND POWERED CAR CARRIER

Before this background a Swedish consortium, consisting of Wallenius Marine, KTH Royal Institute of Technology and SSPA Sweden AB, has begun to design the world's largest sailing ship. The aim is to develop a wind-powered vessel from concept to a technical and financially viable design. This "Wind Powered Car Carrier", called *Oceanbird*, (Figure 1, Table 1) will have five 80-metre-high wing sails targeting emission reductions in the order of 90%. An Atlantic crossing will take around twelve days, compared to seven days in a conventional vessel.

As part of the research and development work for this vessel, seakeeping and manoeuvring tests with a 5-metre-long scale model were conducted in SSPA's Maritime Dynamics Laboratory (MDL), Figure 2.

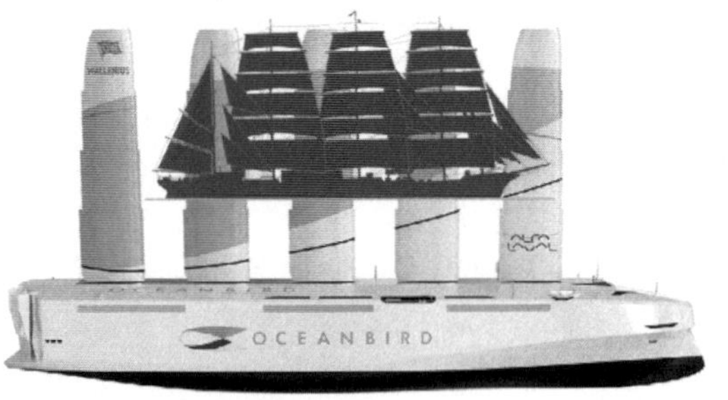

Figure 1: *Oceanbird* and 4-masted barque of *Padua* vintage, approximately to scale.

Table 1: Main Particulars

	Oceanbird	*Padua*
LPP	206.6 m	114.4 m
Beam	39.0 m	14.0 m
Draft	8.5 m	6.8 m
Airdraft	105 m	51.3 m
Volume	30843 m^3	
GM	5.607 m	
Roll gyradius	13.65 m	
Pitch & yaw gyradii	51.65 m	

Figure 2: Wave tests with a model of *Oceanbird*. Note that the wing sails have been replaced by fans/airscrews.

1.3 SEAKEEPING AND MANOEUVRING TESTS FOR SAILING VESSELS

For conventionally propelled merchant vessels like ferries, bulkers, tankers, container vessels etc. an analysis of the seakeeping and manoeuvring properties is firmly integrated in the design process. A few reasons for this are:

- Compliance with IMO manoeuvring standards requires model tests.
- To study ship motions and accelerations for e.g. crew and passenger comfort or structural load calculations.
- To gain an understanding of "added resistance" in waves. This is important in the context of hull form improvements and selecting an engine (sea margin).
- EEDI "minimum power" requirements. IMO requires model testing in regular waves at low speeds.

A study of the above subjects is of course also interesting for the case of a wind propelled ship. Such an investigation needs to include the sail forces because:

- Sails will affect the manoeuvring characteristics.
- If the sails are included or modelled during a test, manoeuvres can be studied experimentally.
- Aerodynamic damping and inertia of the sails affect ship motions directly and added resistance in waves indirectly. An understanding of these effects is required to make power predictions.
- For a constant ship speed, the thrust from the sails will unload the propeller. This reduces the suction on the aftbody (decrease in thrust deduction factor). Such effects need to be considered in any power prediction.
- Course keeping and steering in waves (broaching etc.) can be studied for different sailing and motor-sailing strategies. How does ship behaviour change from 100% motor to 100% sail under otherwise identical conditions?

The purpose of the seakeeping test campaign with *Oceanbird* was to:

- Develop a methodology for model testing wind propelled ships in a realistic environment.
- Collect experimental data that can directly be used to assess ship performance:
    - IMO manoeuvring tests
    - Course keeping and steering in waves
    - Effect of sailing and motor-sailing on course keeping
    - Responses in a seaway (regular and irregular waves)
    - Accelerations in rig due to waves (rig structural design)
    - Added resistance in waves
- Collect data for verification & validation of software developed during project, e.g.:
    - Velocity Prediction Program (VPP)
    - Manoeuvring simulations under sail
    - Predictions of responses and added resistance in waves. (empirical formulations, strip theory, non-linear panel code, CFD)

2. TEST METHODOLOGY

The development of performance prediction and model testing techniques for sailing merchant vessels is still in its infancy. As of today, there are no agreed standards or guidelines how such experiments should be conducted. Worldwide there is only a handful of research facilities that have carried out such tests and they all use different testing methodologies. To address this issue the International Towing Tank Conference (ITTC) has, in June 2021, established a specialist committee on "Wind Powered and Wind Assisted Ships" and it is expected that this committee will review and structure existing model testing techniques.

2.1. THE LITERATURE

While seakeeping and manoeuvring tests are common for conventional cargo vessels, relatively little has been published on such tests for sailing ships. For the reasons discussed above one has to look at the field of yacht design to find relevant information.

Most of the available work from the yachting-sphere focuses on motion responses in waves [3] and added resistance [4], [5]. A review of available methods to study unsteady hydrodynamic behaviour of sailing yachts can be found in Eggers [6]. Broadly speaking, methods can be grouped as follows:

- Captive tests
- Free sailing tests

With the various sub-categories shown in Figure 3.

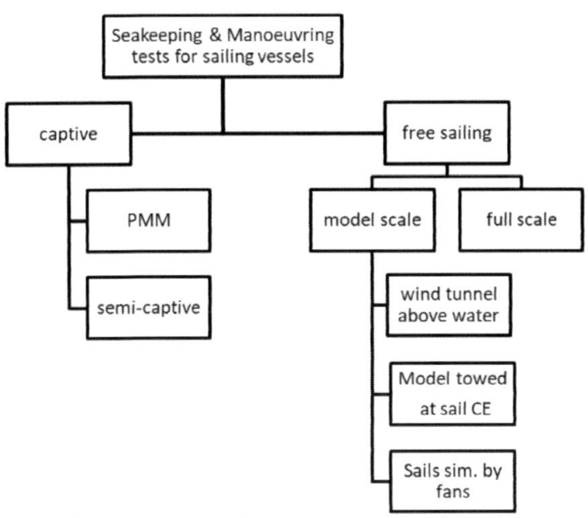

Figure 3: Alternatives for seakeeping and manoeuvring tests with sailing ships.

As discussed by Eggers, the "work horse" of yacht testing is semi-captive testing. In this method a scale model of the yacht, typically free to heave and pitch, but restrained in all other degrees of freedom is towed through the tank and the forces are measured. Such a test setup is used by most research institutes to determine the hull forces in calm water or in waves for various speeds, heeling, drift, and rudder angles. Results, mainly resistance curves and values for added resistance in waves, can be used as input to a VPP. A variation of this technique is the use of a planar motion mechanism (PMM) that forces the model to oscillate in pre-described directions. The results of PMM tests become useful when implemented in a manoeuvring or seakeeping simulation model and can be used to study behaviour associated with rigid body motions and manoeuvring.

Seakeeping and manoeuvring of yachts can also be studied during "free sailing" tests, be it in model or full-scale.

Full-scale tests with sailing yachts are carried out with "sailing dynamometers", purpose-build yachts that can measure the aerodynamic and hydrodynamic forces and moments separately. Such a boat usually comprises a rigid framework structure inside the hull to which all rigging components are attached. The framework structure is connected to the hull shell via force transducers. In this way the aerodynamic loads from the rig can be determined. In a similar way keel and rudder can be equipped with strain gauges to separately measure the forces they exert on the hull. Such a sailing dynamometer was first constructed at MIT, [7] and later in Japan [8] and at the University of Berlin [9]. As summarised by Hansen [10] the focus of such full-scale tests can be to compare measurements to results from CFD simulations or model tests. Similarly results can be analysed to obtain aerodynamic force coefficients of the sails or for CFD validation.

Free sailing tests with models of yachts are relatively rare. The main difficulty here lies in in the application of the aerodynamic forces from the sails.

Equipping the model with scaled-down sails and using fans to blow wind over them is beset by a number of problems. Firstly, it is difficult to create good-quality uniform air flow in a model testing basin. Secondly, scale effects are impossible to handle in such an approach. The hydrodynamic forces require Froude scaling whereas the aerodynamic forces are Reynolds number dependent. Achieving flow similitude for both in the same experiment is next to impossible. Nevertheless, such an approach has been taken by Eggers [11] when a makeshift wind tunnel was installed under the carriage of a seakeeping basin, Figure 4. Experience however, showed that the results might have been contaminated by aerodynamic blockage effects and that the flow quality in stern quartering winds was not optimal.

Figure 4: Simulation of sail forces using a "wind tunnel" above the water surface of a seakeeping basin, from [11].

Another way of applying the aerodynamic forces is to tow the model at the sail centre of effort, see Figure 5 for an example and [6], [12] for more details.
Gauvain [13] reviews a number of similar methods developed over the last 60 years. Although all these methods class themselves as "free sailing" the model is still connected to the tank carriage by means of a towrope or wire. It is unclear how this rope interacts with the model and how model motions are affected, especially in waves and during quick changes of direction.

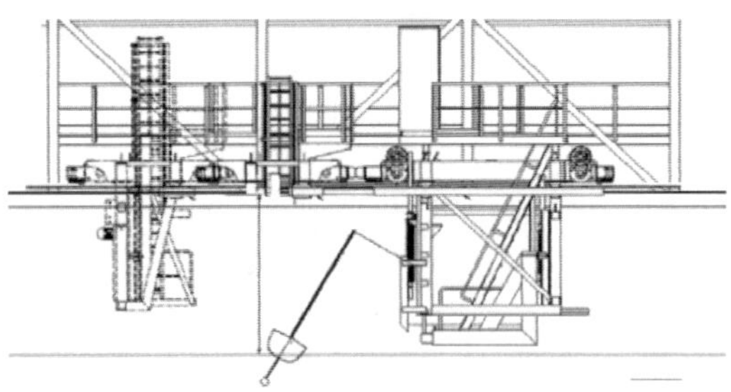

Figure 5: Simulation of sail forces by means of constant tension winch on the carriage of a seakeeping basin, from [6].

A third, relatively new, method of applying the sail forces is to propel the model by a fan (airscrew) mounted on a short mast, Figure 6, [13].
Tsukada et al. [14] report on a similar method to study wind loads on conventionally propelled cargo ship. Here multiple, ducted fans are used to simulate wind forces (drag) acting on the superstructure of conventionally propelled merchant vessels.

Figure 6: Free sailing model tests. Simulation of sail forces using a fan, from [13].

2.2. ACHIEVING SIMILITUDE IN SEAKEEPING MODEL TESTS: A TRILEMMA (AT LEAST)

To obtain meaningful results during any model tests the flow physics need to be modelled correctly. In the case of a conventionally propelled ship this leads to the famous "dilemma of model testing":

1. To model the "wavemaking resistance" correctly the model needs to sail slower than the full-scale ship. Correct scaling requires a reduction of speed by a factor of the square root of the scale factor.
2. To model the "frictional resistance" correctly, "Reynolds scaling" and a corresponding increase in test speed is required.

It is not possible to achieve similitude on both accounts at the same time. The dilemma was famously solved by William Froude by proposing to carry out model tests so that a) the first condition above is satisfied (Froude-scaling) and b) correcting for the resulting error in Reynolds number empirically.
In the case of a sailing ship, and particularly when using the "model sails + wind tunnel" approach from Figure 4 the above dilemma turns into a "trilemma". In addition to 1. and 2. above we get:

3. Aerodynamic similitude requires identical Reynolds numbers of model and full-scale sails.

Not enough: In the case of a model that sails in a seaway, additional unsteady aerodynamic effects might get important which means that the "reduced frequency" of the sails now becomes an additional parameter that needs to be considered [20]. Additionally, rudder and appendage forces due to leeway or manoeuvres might require Reynolds scaling of below-water parts.

Figure 9 shows some of the similarity numbers and their values during a typical, Froude scaled, seakeeping model test. As can be seen the situation is complex.

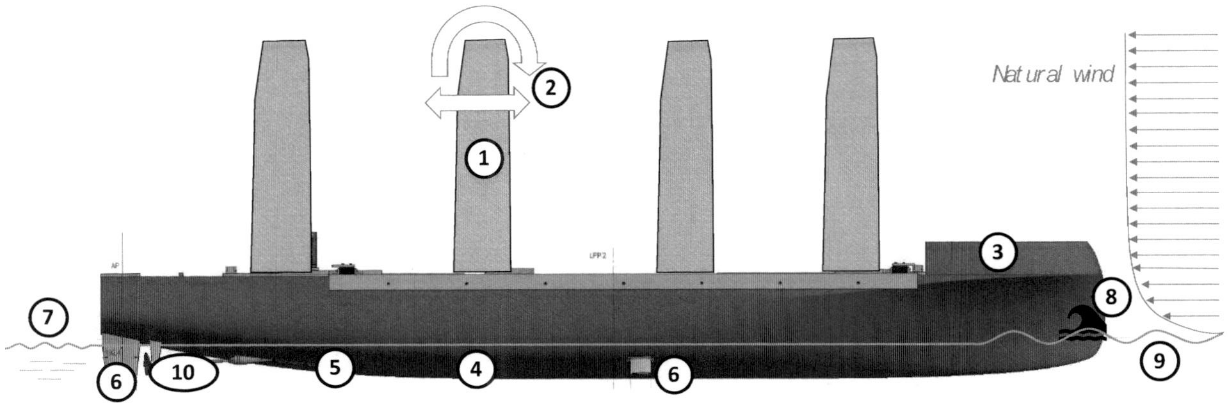

Component	①	③	②	④	⑤	⑥	⑩	⑦	⑧	⑨
	Aerodynamic			Hydrodynamic						
Name	Lift & drag wingsail (steady)	Drag of superstruct.	Lift & drag wingsail (unsteady)	Friction resistance (of hull)	Viscous pressure resistance	Lift & drag appendages & rudder	Propeller forces	Wavemaking resistance	Wavebreak. resistance	Added res. in waves
Source	Lift, induced drag, profile drag	Profile drag = pressure +skin friction	Added mass & unsteady vortex wake	Shear stresses in BL	Pressure distribution	Lift, induced drag, profile drag	Lift, induced drag, profile drag	Generated wave system	Wavebr. near bow	Radiated waves, reflection,
Scaling	Reynolds	Reynolds	Reduced frequency ***k***	Reynolds (hull)	Reynolds (hull)	Reynolds (append.)	Reynolds (prop)	Froude (hull)		
Typical full-scale value	$Re=29\cdot10^6$	$Re=30\cdot10^6$	$k=0.7$	$Re=1088\cdot10^6$		$Re=43\cdot10^6$		$Fn=0.14$		
Model-scale value	$Re<<1\cdot10^6$			$Re=4.1\cdot10^6$		$Re=0.16\cdot10^6$		$Fn=0.14$		

Figure 9: Some of the forces acting on a 1:41.2 scale model of *Oceanbird* and their scaling laws.

2.3. SOLVING THE TRILEMMA: HYBRID TESTING AND SIMULATION OF SAIL FORCES BY FANS

One possible solution to the above trilemma is "hybrid testing". Hybrid testing (sometimes called "software in the loop" testing) is an experimental technique in which the physical system under study is divided into two (or more) subsystems:

- A physical subsystem that is tested experimentally in model-scale
- A numerical subsystem that is simulated on a computer.

The two parts interact in real time using a network of sensors and actuators. Such hybrid testing techniques can be employed when both hydrodynamic and aerodynamic effects are important but cannot be replicated during the same test. Examples include offshore installations such as oil rigs, ocean energy conversion platforms, floating wind turbines and other structures that are affected by both current/waves and wind loads.

Figure 7 shows such a setup for wave-testing a floating wind turbine. Here the wind forces are simulated by a number of winches and cables while the hydrodynamic wave loads on the below-water part of the turbine are produced by the waves in the test basin.

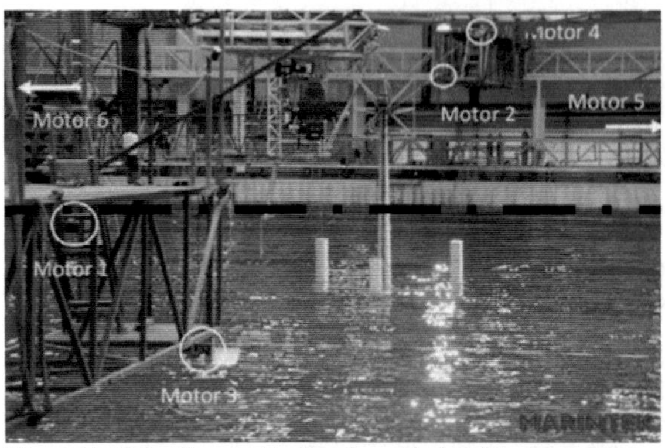

Figure 7: Hybrid test of a floating wind turbine, courtesy of SINTEF Ocean.

Similar approaches have been used for wind tunnel testing yacht sails. Hansen et al. [15]. describe a "Real Time Velocity Prediction Program (RT-VPP)" for wind tunnel testing yacht sails, Figure 8.

Figure 8: Hybrid test of a sailing yacht, from [15].

Here the aerodynamic forces of the sails are measured directly by means of the wind tunnel balance while the hydrodynamic forces acting on the hull are modelled numerically. The RT-VPP then uses sensors and actuators to heel the yacht to the correct equilibrium angles.

The "hybrid testing approach" that was used to investigate the seakeeping and manoeuvring properties of *Oceanbird* is the "mirror image" of Hansen's RT-VPP: Now the sail forces are simulated (by fans) while the hull is tested in waves. A schematic of this setup is shown in Figure 10. The setup consists of:

- The physical subsystem "hull" that is tested in the seakeeping basin
- The numerical subsystem "sails" that is simulated by the two azimuth and *rpm*-controlled fans.

As illustrated, strain gauges are used to measure the axial force of each fan. This allows determining the actual "sail" forces and moments.

In the present case SSPA's inhouse VPP-software "Seaman" was used to determine full-scale "target" sail forces for each tested combination of true wind direction, true wind speed, and ship speed. Calculations are based on lifting line models of *Oceanbird*'s wingsail rig. The details behind the aerodynamic calculations can be found in [16]. The aerodynamic properties of the installed airscrews and scaling considerations are then used to convert these target values into azimuth angle and *rpm*-values for each fan.

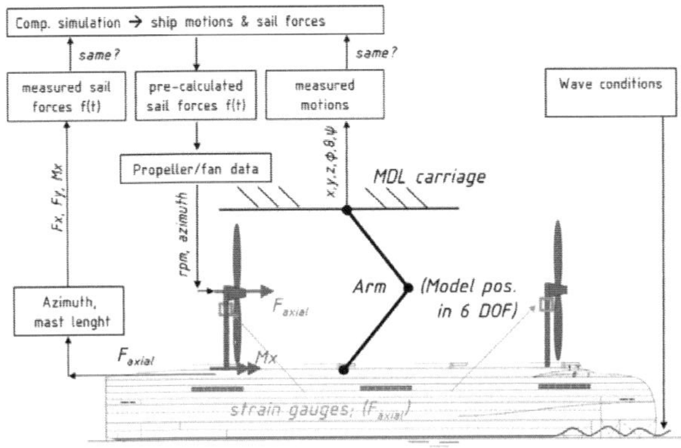

Figure 10: Example of "open- loop" setup used in this paper.

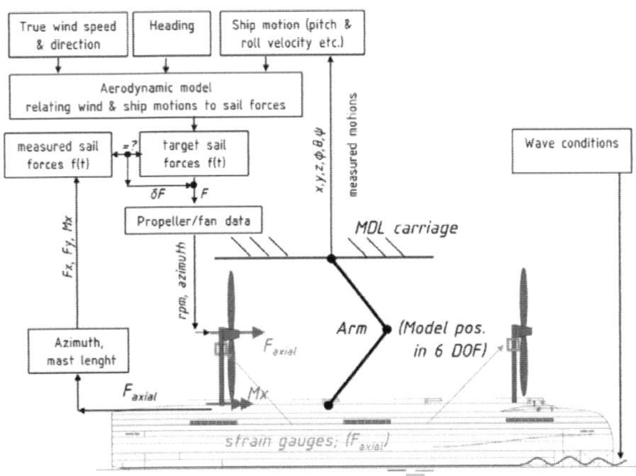

Figure 11: Example of more complicated "close- loop" hybrid setup for future tests.

Figure 11 finally illustrates a more evolved version of the setup from Figure 10. The main difference between the "open-loop" (Figure 10) and "closed-loop" (Figure 11) setups is that in the closed-loop version the aerodynamic "target" forces are not pre-calculated but come from "real-time" evaluation of the model motions and the wind speed and direction. It is envisaged that such a setup can realistically simulate the unsteady behaviour of the sails, including "added inertia" and "vortex shedding" effects [17], [18].
Another difference between the setups is how the "measured sail forces" are treated. In the close-loop version, the difference between measured and target values, δF, is fed-back to the fan-controller and minimised.

2.4. CURRENT SETUP

Figure 12 shows the actual fan-setup used for the tests in SSPA's Maritime Dynamics Laboratory (MDL).
Prior to the tests the *rpm*-thrust relationship of each mast mounted fan was determined during bench tests. The results from these tests (at zero forward speed) were also used during model tests in the basin (at forward speed). This simplified approach could be taken because of the low advance ratio of the airscrews during all basin tests. (low boat speed compared to the circumferential speed of the fans).

Figure 12: Seakeeping model of *Oceanbird*, 1:41.2 scale.

As can be seen, the "masts" of the fans are relatively short. The reason for this is the limited vertical clearance between the model and the carriage of the test facility (Figure 2). As a result, the vertical centre of effort of the fan-forces is about 50% lower than the centre of effort of the wing sails. The resulting error in heeling moment and angle was estimated and found to be negligible because of the large initial stability of the hull.

2.5. SCALING OF SAIL FORCES

The hybrid testing approach is based on calculating full-scale "target sail forces" and replicating these with model mounted fans. In this method the aerodynamic properties of the installed airscrews and "scaling-ratios" are then used to convert these (full-scale) target values into (model-scale) azimuth angle and *rpm*-values for each fan. This approach poses one major challenge: How to scale the aerodynamic forces from full-scale to model-scale?
Especially the scaling of aerodynamic side forces (from the sails) in comparison to the hydrodynamic forces (from the appendages) turned out to be critical during the experimental campaign. The issue became apparent during the first free-sailing tests when the leeway and rudder angles converged at unrealistic values that were much higher than predicted by the VPP, which modelled the full-scale problem. The most likely reason for this is that the model rudders operate at too low Reynolds numbers compared to full-scale. To achieve similitude of wave-induce forces during seakeeping tests the model-speed needs to be Froude-scaled. As shown in Figure 9 this results in Reynolds numbers of the appendages that are two orders of magnitude too low. To make matters worse, the full-scale values, $Re \approx 40 \cdot 10^6$, are supercritical (i.e. above the value of $Re \approx 1 \cdot 10^6$ where lift and drag

characteristics of lifting surfaces become Re-independent) whereas the model rudders operate at approximately Re ≈ $0.16 \cdot 10^6$, i.e. in a range where lift and drag drastically change their behaviour [19], [20]. As a result of this Reynolds-number "mismatch" it can be expected that, at identical rudder angels, the model rudder will create a side force that is lower than the one of its full-scale counterpart, see [21] for a more detailed discussion. The same problem of course occurs during model tests with conventionally propelled ship models but is somewhat mitigated by the fact that the rudder operates in the propeller "race". As ITTC recommended procedure and guideline 7.5-02-06-01 (free-running model tests) puts it: "Since the rudder(s) are normally positioned in the wake field behind the ship and in the propeller race, i.e., in a very disturbed and turbulent flow, the Reynolds number effect for the rudder forces may be neglected. Nevertheless, sand strips or studs are sometimes applied to the rudder."

On sailing ships and in the absence of propellers, the Reynolds-number mismatch gets more important. During experiments with sailing yachts the effects are usually minimised by testing as large a model as possible in combination with "turbulence stimulation" on the appendages [22]. The use of "trip-wires", studs, zig-zag tape, etc. however results in another problem: The "stimulators" create drag that somehow needs to be accounted for.

In the current project an alternative way to minimise the effect of the low Reynolds numbers of the model rudders was investigated. Figure 13 illustrates several possible strategies how the Reynolds number related mismatch of side forces can be tackled. The "orange" approach"on the left corresponds to the classical method of simulating a higher Reynolds number by testing in the propeller race or by stimulating turbulence. In the middle/"grey" approach the rudder is enlarged to compensate for Re-related loss in lift. In this paper the "green" approach shown on the RHS is explored i.e., the aerodynamic fan-forces, particularly the side force, were reduced to compensate for the loss of rudder side force. In practice this was achieved by adjusting the aerodynamic side force from the fans until the observed rudder- and leeway angles corresponded approximately to the full-scale predictions. After the test campaign the experience and findings from such ad-hoc/ trial & error adjustments were systematically analysed, and this resulted in the sub-approaches A, B, and C in Figure 13. Approaches A and B were tried during the experimental campaign, resulting in a suggestion for future tests (approach C).

2.5. (a) Approach A

Scaling of aerodynamic drive force in model-scale (Fx_M) until the model speed reaches its target value. This determines a "resistance based" scaling ratio b:

$$\beta = \frac{Fx_M}{Fx_S}$$

Scaling of aerodynamic side force and yawing moment with same factor.

$$Fy_M = \beta \cdot Fy_S$$

$$Mz_M = \alpha \cdot \beta \cdot Mz_S$$

With a geometric scale factor.

Because the leeway and rudder angles resulting from this approach were unrealistically high, approach B was developed.

2.5. (b) Approach B

As approach A but introduction of an additional scaling ratio g_1 to account for the Reynolds-related reduction in rudder and appendage side force:

$$\beta = \frac{Fx_M}{Fx_S}$$

$$Fy_M = \gamma_1 \cdot \beta \cdot Fy_S$$

$$Mz_M = \alpha \cdot \beta \cdot Mz_S$$

Where g_1 was found by "trial and error" adjustments until the observed rudder- and leeway angles corresponded approximately with the full-scale predictions. This approach keeps the yawing moment at the value from approach A) and reduces the total aerodynamic side force. Experimental results now looked realistic. After the experimental campaign approach B was modified into approach C).

2.5. (c) Approach C

Similar to approach B but instead of reducing side force the yawing moment is adjusted via g_2:

$$\beta = \frac{Fx_M}{Fx_S}$$

$$Fy_M = \beta \cdot Fy_S$$

$$Mz_M = \gamma_2 \cdot \alpha \cdot \beta \cdot Mz_S$$

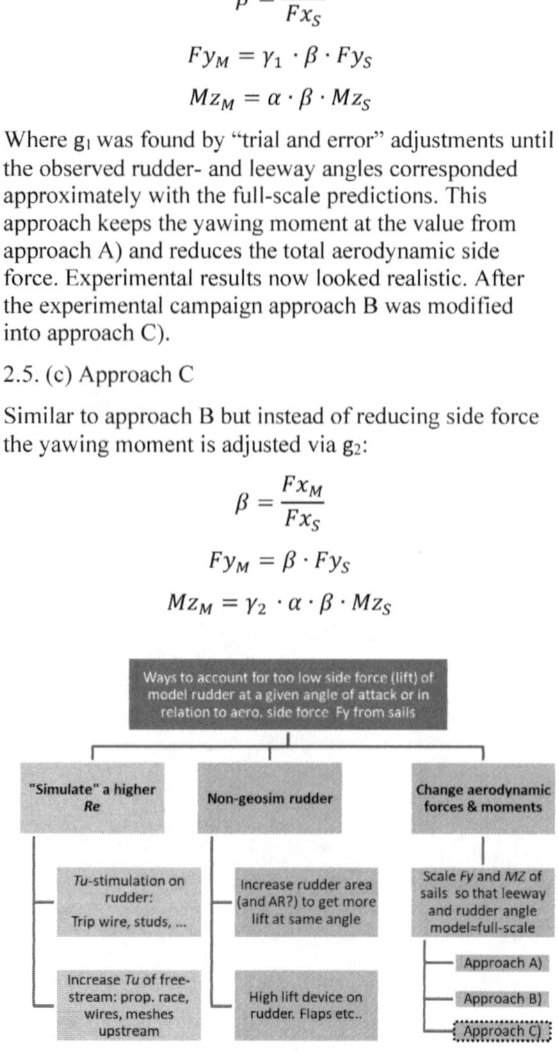

Figure 13: Different strategies to account for Reynolds number mismatch of flow around rudders during model tests.

2.6. FAN FORCES CONTROL

The target forces for the fans were calculated with SSPA's inhouse VPP-software "Seaman" and are based on lifting line calculations. The *rpm*-thrust calibrations of the airscrews (from bench tests) and the geometry of the setup are used to convert these target values into azimuth angles (*A1 A2*) and *rpm*-values for each of the two fans. The fans then produce the "forces *F1* and *F2*, see Figure 14.

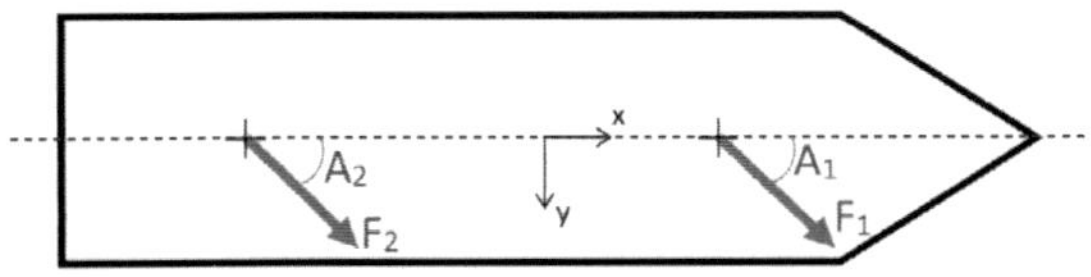

Figure 14: Fan force and angle definitions.

The sail forces and their x- and y components ("driving force" and "side force") depend i.a. on the ship's course relative to the "true wind". To correctly replicate this behaviour, the fan forces and angles were adjusted in real-time during the tests. To this end the target aerodynamic forces from the VPP were calculated for a range of "true wind angles" (TWA) and polynomials used to interpolate between them.

Figure 15 shows a plot of these polynomials. In this case it was decided that *F1* should equal *F2* and that the target yawing moment, M_z should be achieved by adjustment of *A1* and *A2*. For the case shown in the figure:

True wind speed, *TWS*=24.6 kts
No waves (artificial test case)
Ship speed, V_s=12 kts
F1 := *F2*

The bottom part of the figure shows the same data but converted into drive and side force. As expected, the side (heeling) force is largest when sailing close hauled (TWA around 60°) while the driving force from the rigid wing sails has a flat maximum for true wind angles of around 90°.

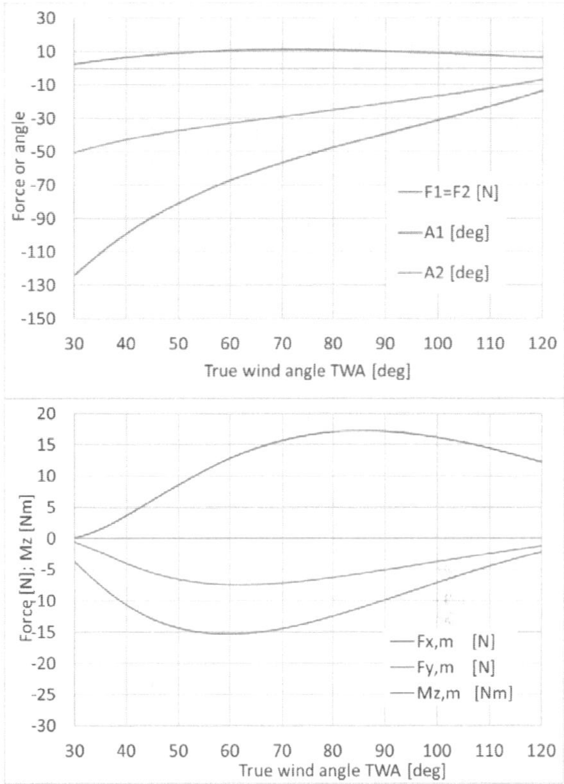

Figure 15: Polynomials used to model sail forces. Fan forces and angles (top). Same data converted to total drive and side force produced by both fans together (bottom).

2.7. PROPELLERS DURING SAILING TEST

For the sake of efficiency, the rigid, fixed pitch propellers remained on the model during the entire test campaign. In the cases where the ship was 100% sailing the propeller *rpm* was adjusted in such a way that the measured thrust on the propeller shaft was close to zero. It was found that such "slowly rotating" propellers represent the propellers envisaged for *Oceanbird* relatively well.

3. SELECTED RESULTS

A selection of results from the experimental campaign follows below. All tests were conducted in SSPA's Maritime Dynamics Laboratory (MDL). This basin has a size of 88 m x 39 m and a maximum water depth of 3 m. A multi-motion carriage for model control and data acquisition spans the width of the basin, Figure 2. See Kim et al. for more details [23].

3.1. IMO MANOEUVRING TESTS

Example results from standard IMO zig-zag tests according to IMO MSC/Circ.1053 and MSC.137(76) are shown in Figure 16. These tests were conducted for both, engine and sail propulsion. For all cases the speed was 12 knots and the rudder turning rate 2.32 °/s (full-scale). The manoeuvring tests in sailing condition were carried out in "beam winds" from the port side.

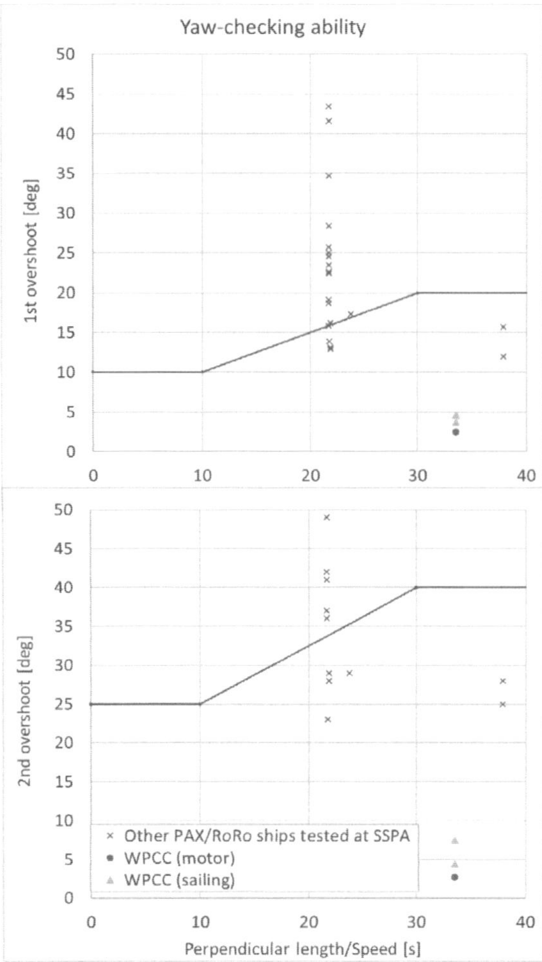

Figure 16: Results of 10/10 zig-zag tests.

As can be seen, the yaw checking ability of *Oceanbird* with its twin rudders is excellent. Although increasing under sail, overshoot angles are still from the IMO limits.

3.2. RESPONSES IN A SEAWAY

Figure 17 illustrates a typical seakeeping tests under sail. Such raw data was then analysed to study e.g. course keeping in waves or added resistance in waves, see below.

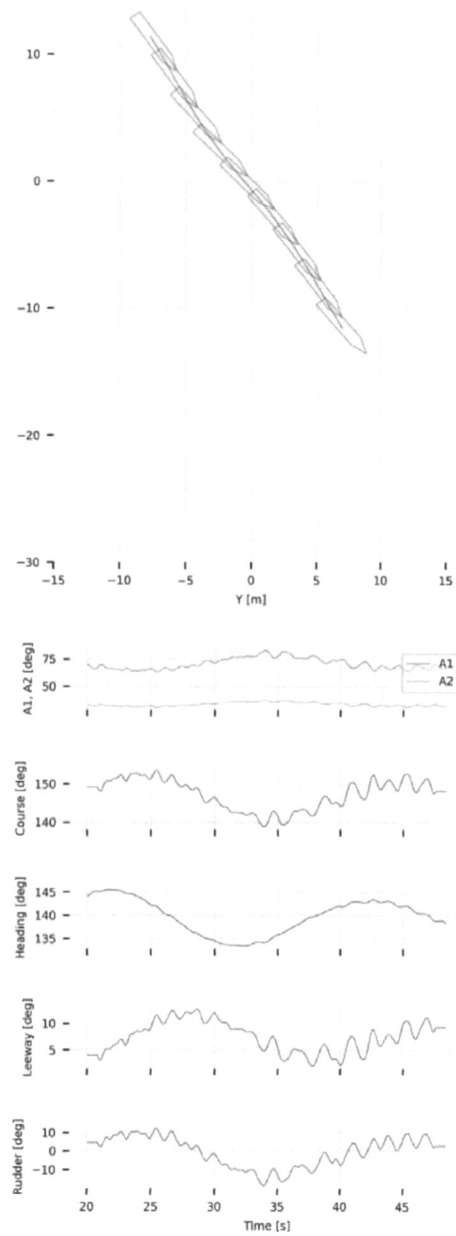

Figure 17: Results from seakeeping tests at 12 knots. Sail propulsion, Significant wave height $H_s = 3.25$ m, T_z: 6.90 s, Wave direction: -55° off the bow.

3.3 COURSE KEEPING AND STEERING IN WAVES

Results of the type shown in the preceding section were, i.a., analysed to study course keeping in waves and how this is affected by sailing. Figure 18 summarises some results. Shown are course deviations and rudder angle values for true wind angles of 55° and 120° in calm water and two sea states (SS5 and SS7) for both, sail and motor propulsion. As can be seen, course deviations increase somewhat under sail but are still small indicating that the steerability for the tested conditions is good.

As far as rudder angles are concerned, it can be observed that rudder action (standard deviation) increases with wave height and is larger for the sailing cases. The most likely reason for this is the absence of the "propeller race" in the sailing cases and the resulting loss of rudder efficiency.

For some of the sailing cases, mean values of the rudder angle are large (e.g. 17.6° for TWA=120°, SS7), this is probably caused by bad "sail trim" and could be avoided by distributing the sail forces differently between the two fans.

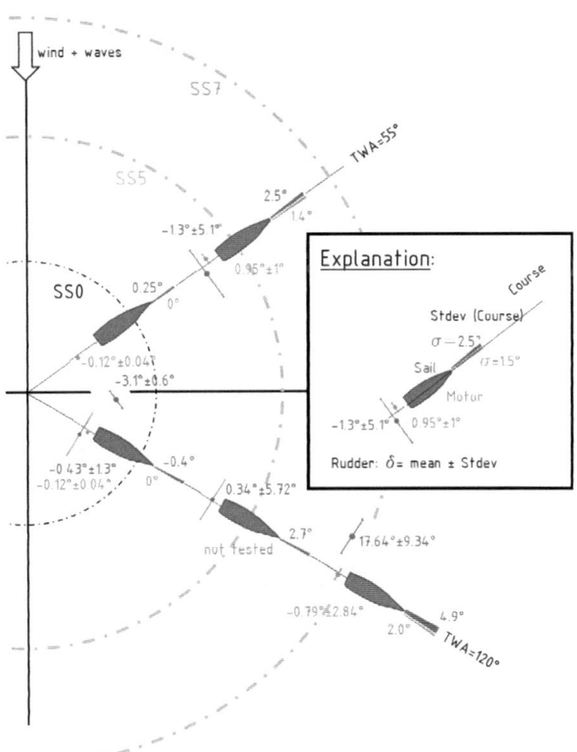

Figure 18: Course keeping in waves. V_s= 12 kts, SS5 (H_s = 3.25 m, T_z 6.90 s) and SS7$_{low}$ (H_s = 6.00 m, T_z = 10.20 s).

3.4 ADDED THRUST IN WAVES

How sailing affects added resistance and added thrust in waves is of course also of interest. Investigations are still ongoing, but it appears that for the sailing cases added thrust values are higher than for motoring, Figure 19. Such information is vital for correct performance prediction.

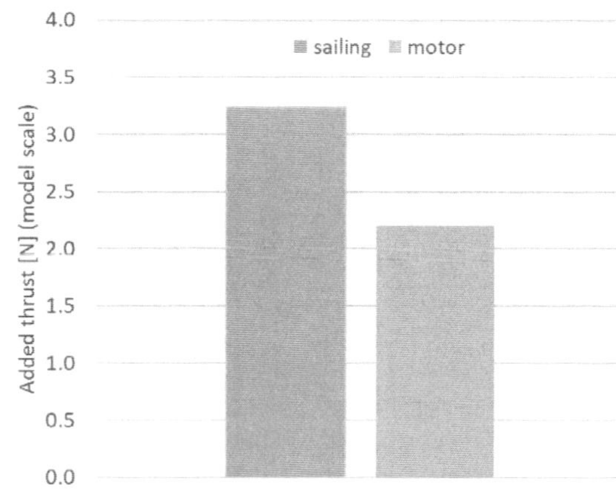

Figure 19: Influence of sailing on added thrust. V_s = 12 kts, SS5 (H_s = 3.25 m, T_z 6.90 s).

Figure 20 compares *Oceanbird* to other ships of similar size. As can be seen, added thrust values for stern quartering waves are average, while values for bow quartering waves are quite low. The fine entry and sharp stem of *Oceanbird* in combination with the absence of a bulb will play a role here.

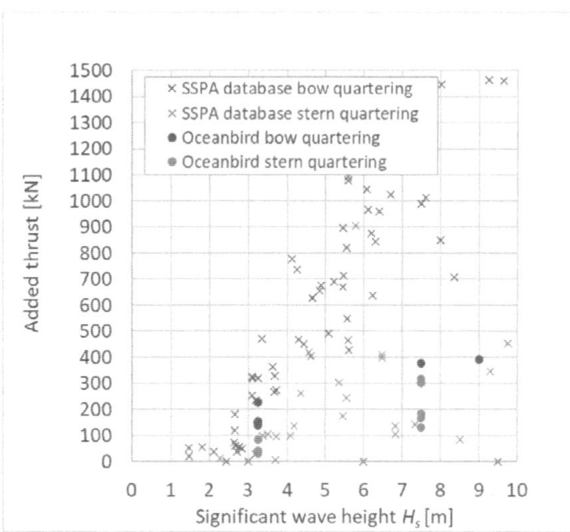

Figure 20: Comparison of added thrust in waves to selected RoRo and RoPax.

3.5. ACCELERATIONS IN THE RIG

Another possible use for the experimental results is the analysis of wave-induced accelerations in the rig. Figure 21 shows an example where the measured accelerations from the seakeeping tests were extrapolated to all sea states and combined with world-wide wind statistics to yield a first estimate of fatigue loads.

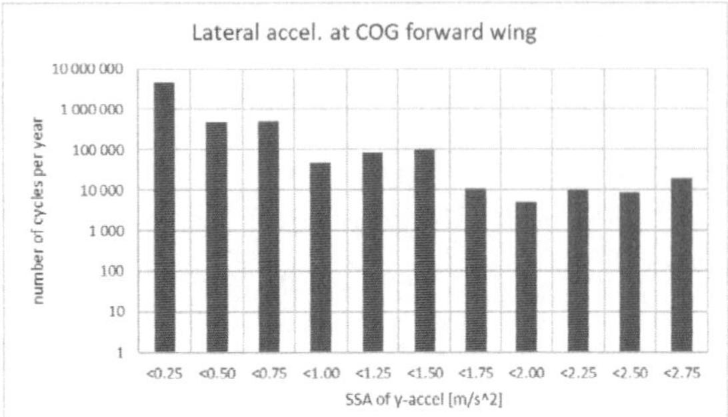

Figure 21: Significant single amplitudes (SSA) of wave induced accelerations at centre of gravity of forward wing.

3.6. VERIFICATION & VALIDATION DATA

The test campaign also resulted in data that can be used to verify and validate software and prediction tools.

3.6. (a) Velocity Prediction Program and route simulations

During the development of the design concept for *Oceanbird,* a large number of VPP-runs and routing simulations in waves were caried out [24]. In most of these, superposition theory and the "STAWAVE-2" model [25] were used to determine added resistance in waves. Figure 22 compares predictions to measurements from MDL. As can be seen, agreement is quite good and all simulations based on the STA model should be on the conservative side (higher added resistance because of higher RAO values).

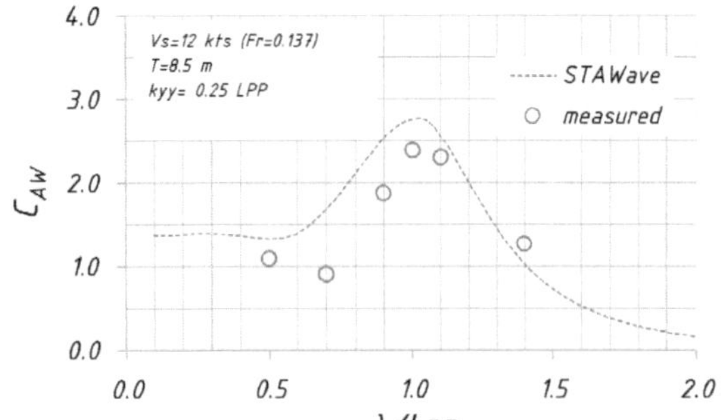

Figure 22: Added resistance transfer function (RAO).

3.6. (b) V+V of seakeeping codes

Another potential application of the experimental results is the validation of numerical seakeeping analyses. Figure 23 shows an example of a simulation with the non-linear panel code Shipflow Motions. Investigations are still ongoing, but it is planned to validate these simulations against MDL-results.

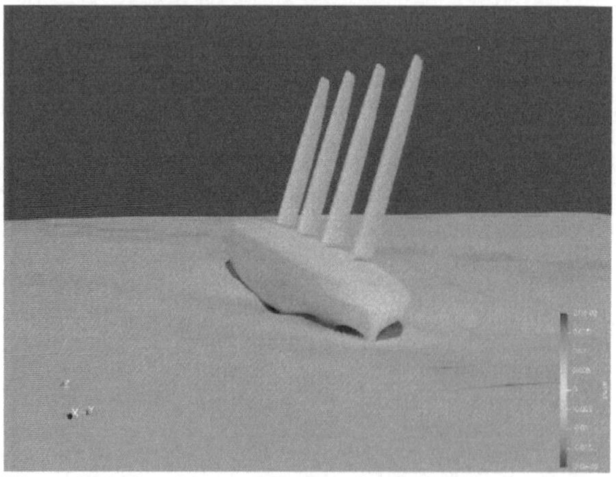

Figure 23: Coupled aero-hydro simulation in the non-linear seakeeping code Shipflow Motions.

4. CONCLUSIONS

The following conclusions can be drawn from the investigation:

4.1. TEST METHODOLOGY

4.1. (a) General findings

The single most important finding from the test campaign is that conducting seakeeping and manoeuvring tests with the open-loop setup shown in Figure 10 is feasible and produces credible results. For the reasons discussed above this "fan-approach" seems to be preferable over both, the "wind tunnel- approach" and the simulation of sail forces by winch and wires.

The encouraging experiences with the open-loop setup seem to warrant a future attempt to build a closed-loop setup with force feedback, Figure 11. Such a setup could not only model steady state sail forces but also simulate the unsteady behaviour of the sails, including "added inertia" and "vortex shedding".

4.1. (b) Rudder scale effect

One of the areas that requires further investigation is how to deal with the unavoidable Reynolds number effects on the rudders of a free sailing wind propelled ship model. Here the experimental campaign resulted in important insights and a number of strategies to minimise scale effects were investigated. Suggestions for future tests include approach C from Figure 13. Alternatively, the idea of using non-geosim rudders (middle track in Figure 13 seems to be worth exploring.

4.2. SHIP CONCEPT

As far as *Oceanbird* is concerned the following conclusions can be drawn for the limited number of conditions tested during the campaign:

- IMO manoeuvring criteria will very likely be met under power and at a speed of 12 knots.
- Criteria derived from zig-zag tests (initial turning, yaw-checking and course-keeping ability) will even be met under sail alone and in moderate beam winds.
- The added resistance of *Oceanbird* in head and bow waves is small compared to other RoRo/RoPAx vessels from SSPA's database.
- The effect of sailing on added resistance requires further investigation.
- For true wind angles of 55° and 120°, course keeping and steering in waves up to 6 m significant height should not be an issue. Due to the absence of the propeller race rudder action will be somewhat larger than for motoring under the same conditions. Trimming the sails will make it possible to minimise rudder angle mean values.
- No test in following seas and at low speeds were conducted. Consequently, the risk of broaching could not be assessed.
- Future investigations should include such broaching tests and, depending on hull form, a study of "parametric roll". While the roll damping of the closely sheeted sails will almost certainly mitigate the risk of parametric roll in upwind conditions, downwind conditions (with sheeting angles close to 90° and little aerodynamic roll damping) could still be critical.

5. REFERENCES

[1] GLAUERT, H., 'The elements of aerofoil and airscrew theory', Cambridge, 1926.

[2] T. CHOU, V. KOSMAS, K. RENKEN, M. ACCIARO, 'New Wind Propulsion Technology - A Literature Review of Recent Adoptions,' Interreg North Sea Europe Wind Assisted Ship Propulsion (WASP) report D 5.B, 2020.

[3] OTTOSSON, P., BROWN, M., LARSSON, L., 'The Effect of Pitch Radius of Gyration on Sailing Yacht Performance,' *High Performance Yacht Design Conference*, 2002.

[4] SUTCLIFFE, C.J., MILLWARD, A., 'The Effect of Pitch Moment of Inertia in Body Axes on the Performance of a Yacht in Waves,' *Proc. 13th Chesapeake Sailing Yacht Symposium*, 1997.

[5] SKINNER, G.T., 'Sailing vessel dynamics: investigations into aero-hydrodynamic coupling,' *MSc thesis, Massachusetts Institute of Technology*, 1982.

[6] EGGERS, R., GAILLARDE, G., KONING, J., 'A review of unsteady hydrodynamic behaviour of sailing yachts and methods to study it,' *4th High Performance Yacht Design Conf.*, 2012.

[7] MILGRAM, J. H., PETERS, D. B., ECKHOUSE, D. N., 'Modelling IACC Sail Forces by Combining Measurements with CFD,' *11th Chesapeake Sailing Yacht Symposium*, 1993.

[8] MASUYAMA, Y., FUKASAWA, T., 'Full Scale Measurement of Sail Force and the Validation of Numerical Calculation Method,' *13th Chesapeake Sailing Yacht Symposium*, 1997.

[9] BRANDT, H., HOCHKIRCH, H., 'Full-Scale Hydrodynamic Force Measurements on the Berlin Sail- Force-Dynamometer,' *14th Chesapeake Sailing Yacht Symposium*, 1999.

[10] HANSEN, H., 'Enhanced wind tunnel techniques and aerodynamic force models for yacht sails,' *PhD thesis, Univ. of Auckland*, 2006.

[11] EGGERS, R., KISJES, A.S., 'Seakeeping and manoeuvring for wind assisted ships,' *International Conference on Wind Propulsion*, 2019.

[12] EGGERS, R., 'Prediction methods for Team Akzo Nobel in the Volvo Ocean Race and Wind Assisted Ship Propulsion for merchant vessels,' *HISWA Symposium*, 2018.

[13] GAUVAIN, E., 'The Un-restrained Sailing Yacht Model Tests – A New Approach and Technology Appropriate to Modern Sailing Yacht Seakeeping,' *23th Chesapeake Sailing Yacht Symposium*, 2019.

[14] TSUKADA, Y., SUZUKI, R., UENO M., 'Wind loads simulator for free-running model ship test,' *International Conference on Offshore Mechanics and Arctic Engineering - OMAE*, 2017.

[15] HANSEN, H., JACKSON, P. S., HOCHKIRCH, H., 'Real-Time Velocity Prediction Program for Wind Tunnel Testing of Sailing Yachts,' *The Modern Yacht Conference*, 2003.

[16] MALMEK, K., DHOME, U., LARSSON, L., WERNER, S., RINGSBERG, J., FINNSGÅRD, C., 'Comparison of Two Rapid Numerical Methods for Predicting the Performance of Multiple Rigid Wing-Sails,' *Innovsail*, 2020.

[17] GERHARDT, F.C., FLAY, R., RICHARDS, P., 'Unsteady aerodynamics of two interacting yacht sails in two-dimensional potential flow,' *J. Fluid Mech.*, vol. 668, pp. 551–581, 2011.

[18] GERHARDT, F.C., FLAY, R., RICHARDS, P., 'Unsteady aerodynamic phenomena associated with sailing upwind in waves,' *3rd High Performance Yacht Design Conference*, 2008.

[19] SCHMITZ F.W., 'Aerodynamics of the Model Airplane', *NASA RSIC Report 721*, 1967.

[20] HÖRNER, S.F., BORST, H.V., 'Fluid-Dynamic Lift', 1975.

[21] MARIMON GIOVANNETTI, L., GERHARDT, F.C., KJELLBERG, M., ALEXANDERSSON, M., 'The art of model testing : Using CFD to adapt traditional tank testing techniques to a new era of wind propelled shipping,' in International Conference on Numerical Methods in Marine Engineering, MARINE, 2021.

[22] P. H. Miller, 'Dynamic lift coefficients for spade rudders on yachts,' in 18th Chesapeake Sailing Yacht Symposium, CSYS, 2007, no. March, pp. 223–232, doi: 10.5957/csys-2007-017.

[23] LEE, J., KIM, Y., KIM, B., GERHARDT, F. C., 'Comparative study on analysis methods for added resistance of four ships in head and oblique waves', Ocean Engineering, vol. 236, 2021.

[24] WERNER, S., KUTTENKEULER, J., HÖRTEBORN, A., LUNDBÄCK, O., DHOMÉ, U., 'Route Evaluation Methods for long-distance sailing vessel performance predition', *High Performance Yacht Design Conference*, 2021.

[25] ITTC Recommended Procedure and Guideline 7.5-04-01-01.1, 'Preparation, Conduct and Analysis of Speed/Power Trials', 2021.

6. AUTHORS BIOGRAPHY

Frederik C. Gerhardt is Senior Technical Specialist for seakeeping & ship performance in waves and aerodynamics at SSPA. He is responsible for model tests in the Maritime Dynamics Laboratory and chairs the 30th ITTC Seakeeping Committee.

Martin Kjellberg is Senior Technical Specialist for seakeeping & ship performance in waves at SSPA and is the main developer of SHIPFLOW MOTIONS at Flowtech International AB.

Ida Wigren is Project Manager at SSPA. Her main focus areas are manoeuvring and seakeeping model tests.

Sofia Werner is responsible for method development within fluid dynamics at SSPA. Her previous experience includes towing tank testing and ship design using CFD. She chairs the ITTC Specialist Committee for Wind Powered Ships.

Mikael Razola is the technical director for the Oceanbird Sailtech project at Wallenius and project manager of the Wind Powered Car Carrier research project since 2019, working closely with SSPA Sweden AB and KTH Royal Institute of Technology on various research topics related to sailing ships.

INTERACTIONS BETWEEN TWO FLETTNER ROTORS USED FOR WIND SHIP ASSISTED PROPULSION

B Charrier, BC Consultant, France

SUMMARY

Wind tunnel tests are carried out at the Aerodynamics Laboratory of the Ecole Nationale des Arts et Métiers (ENSAM) in Paris on 50 mm and 130 mm diameter rotating circular cylinder in infinite aspect ratio or finite aspect ratio of 3 to 6, with or without rotating endplates. Aerodynamic forces are measured by a 5-components balance and by integrating pressures over three sections along the span of the rotating cylinder.

The tests carried out at Reynolds numbers between 13,500 and 70,000 are significant and make it possible to dispense tests at high flow speeds.

The results, based on pressure measurements, indicate that the rotation of the cylinder drives the fluid in a spiral motion along the span, winding the flow, accompanied by a very strong increase in the drag coefficient between the middle and the top of the cylinder.

The interaction between two rotating cylinders, spaced out by 6 diameters, decreases the ratio C_L / C_D of the thruster reducing its propulsive performance at close wind angles (30° <β <70°).

NOMENCLATURE

C_D	Overall drag coefficient, D / ½ $\rho V^2 S$
C_L	Overall lift coefficient, L / ½ $\rho V^2 S$
C_p	Pressure coefficient, $(P - P_0)$ / ½ $\rho V^2 S$
D	Overall drag force
d	Diameter of the cylinder
d_e	Diameter of endplate
h	Span of the cylinder
k	Velocity ration, U / V
L	Overall lift force
P	External pressure measured on the rotating cylinder
P_0	Static pressure
R_e	Reynolds number, V.d / ν
S	Cylinder surface area : h.d
U	Cylinder tangential velocity
V	Speed of the upstream infinite flow
ν	Kinematic viscosity of air
ρ	Density of air under the temperature and pressure conditions of the experiment
λ	Aspect ratio : h/d
λ_e	Effective Aspect Ratio : according expérimental configurations $\lambda < \lambda_e < \lambda\infty$

1. INTRODUCTION

To respond to the challenge posed by climate change, scientists have been mobilized for many years. Innovative solutions are proposed to drastically reduce greenhouse gas emissions, mainly CO_2, and methane, in all economic sectors activity sectors of economy. Commercial shipping, which contributes around 3% of global emissions is of particular concern.

In 2017, the International Maritime Organization pledged to reduce CO_2 emissions by 50% in 2050 compared to 2000.

The oil barrel price influences the cost of the ton of goods transported, and therefore the profitability of ships. In the short to medium term, fluctuations in the price of a barrel of oil are unpredictable, but economists are sure that in the future, the price of oil will rise and remain high, at more than $100. The energy transition has also started for commercial shipping.

To meet these challenges by limiting greenhouse gas emissions and reducing the consumption of ship engines, ship owners, charterers, shipyards, engine manufacturers, universities and research centers are mobilized. To reduce the fuel consumption of ships, many tracks are proposed, such as reduction of their speed, optimization of the routes followed, increase in the specific power of diesel engines, GNL or even hydrogen as alternative fuel, optimization of hull shapes, and superstructures, reduction of the frictional resistance of the hull by the emission of air bubbles and obviously wind propulsion. As the maturity of development is not the same for all technologies, research and investment priorities must be decided upon and supported by professionals and public authorities.

Following the two oil crises of 1973 and 1979, scores of researches on wind propulsion for ships were carried out and ships were fitted with efficient systems, mainly in Japan and some in Europe. From 1985, with the return of inexpensive petroleum, wind propulsion for ships was again forgotten. Research on the Flettner rotor has recently resumed and commercial vessels are being re-equipped. This publication deals with the aerodynamic study of rotating cylinders and its application to wind propulsion.

2. BIBLIOGRAPHY

Around 1851, the physicist G. Magnus (1) was interested in the effects caused by the action of the wind on rotating cylinders placed perpendicular to the fluid. His attention was drawn to the influence of the rotation of the projectiles on their deflection, that is to say their lateral displacement in relation to the firing, a problem which faced ballisticians. G. Magnus found that when a cylinder rotates around its axis, the reaction exerted by the fluid is deflected and the magnitude of this deflection varies with the speed of the fluid and with the speed of rotation of the cylinder. It is the dissymmetry of the fluid flow, therefore the distribution of the pressure field around the rotating cylinder, that gives rise to this reaction.

Lord Rayleigh (2) in 1877 proposed the first mathematical model of flow, representing it as a potential flow around a cylinder with circulation.

A. Thom (3) from 1925 to 1930 carried out a comprehensive study of the Magnus effect. He measured the aerodynamic forces on a 5.7 aspect ratio cylinder. He found that the rotation of the cylinder generated very significant aerodynamic forces. Based on the projected surface of the cylinder, the dimensionless aerodynamic coefficients, lift and drag coefficients, were far greater than those obtained on airplane wings. He also found that the cylinder surface state, smooth or rough, had no influence on the aerodynamic characteristics. To approach the two-dimensional flow, cylinders of aspect rations of 12.5 and 26 were fitted with rotating endplates to cylinder ratios of De/D=3. He also observed that the aerodynamic friction torque was almost the same when the cylinder was rotating in a still atmosphere or with wind. This observation will be confirmed later by many authors, including Badalamenti (10), and Bordogna (11).

Ray W. van Aken and H.R. Kelly (4) in 1957 looked at the influence of the Reynolds number on the aerodynamic behavior of the rotating cylinder. They were the first to demonstrate the importance of the critical Reynolds number, defined as the Reynolds number at which the cylinder drag coefficient drops sharply.

Swanson (5) in 1961 was interested in the action of the fluid on a rotating cylinder in infinite ratio by measuring the aerodynamic forces on the central part of the cylinder in order to get a result independent of the end effects. He explains the reversal effect of the Magnus effect observed at low values of U/V, by the length of the boundary layer on either side of the starting point of this boundary layer.

In 1979, Bertrand Charrier (6), author of this article, carried out a theoretical and experimental study on the Magnus effect for the use of ship propulsion. The very sharp increase in the price of a barrel of oil, a consequence of the two oil crises of 1973 and 1979, stimulated the search for efficient, compact, low-maintenance wind thrusters, adapted to the easements of modern ships allowing energy savings in commercial shipping. The intuition of Prof. Lucien Malavard, thesis Director, member of the French Académie des Sciences, and an aerodynamic high level scientist, was that wind would once again become a source of energy for ships. Indeed, Anton Flettner (7), German engineer had applied the Magnus effect for the propulsion of ships. He fitted the Buckau, a 52 m vessel with two vertical cylinders of 15 m high and 2.75 m in diameter fitted with small rotating endplates. In 1926 Flettner embarked on the Hamburg-New York crossing with his ship then renamed BADEN-BADEN. The Hamburg shipping company Amerika Line then made an order for ten rotor ships. But only one ship was built. 92 m long and equipped with 3 rotors (h = 17 m, Ø = 4 m, rotation speed = 150 rpm), the BARBARA sailed on the Hamburg-Italy line, carrying 3,000 t of goods. After six years of use, the cylinders were dismantled and the ship was fitted with diesel propulsion. The work of B. Charrier made possible to quantify the aerodynamic performances of rotating cylinders of 2 to 6 aspect ratio, fitted or not with rotating endplates, to show that the performances were practically independent of the Reynolds number when U/V is greater than 1.5-2, to study the interaction between two rotating cylinders and to highlight the twisting of the flow along the span of the rotating cylinder. This lead to favor the overall measurements of the forces rather than the pressure measurements.

Drawing lessons on the performance of the rotating cylinder, in particular on the energy expended to drive it, B. Charrier (8) studied with Pr. L. Malavard (9) very thick profiles whose aerodynamic performance was equivalent to those of the rotating cylinder, but required less energy to create the propulsive effect. In 1985, the Alcyone ship of the Cousteau foundation was fitted with two cylinder-flap wings with boundary layer suction, called Suction Sails and Turbosail as commercial name by Cousteau Foundation.

C. Badalamenti and S. A. Prince (10) studied in 2008 the effect of endplates of different sizes on the aerodynamic performance of the rotating cylinder. They confirmed that the power required to rotate the cylinder was independent of the Reynolds number.

The work of C. Bordogna (11) in 2019 on a rotating cylinder of aspect ratio of 3.73 placed between 2 walls, simulating the infinite ratio and without endplates, show that the lift and drag coefficients are independent of the Reynolds number for U / V greater than 1.5 - 2.0. This work confirms the results published by A. De Marco (12), and Badalamenti (10).

Since 2010, several commercial ships have been fitted with rotating cylinders. The Finnish company Norsepower has equipped several ships with large rotation cylinders. Sea trial tests were carried out on the Maersk Pelican tanker, equipped with two 30 m high rotating cylinders, achieving 8.2% fuel savings over one year of operation (13).

3. EXPERIMENTATION

3.1. WIND TUNNEL CHARACTERICS

The aerodynamic tests were carried out in the wind tunnel of the Ecole Nationale Superieure d'Arts et Métiers in

Paris. The stream is of the semi-guided type with a rectangular frontal section of 1.65 m x 1.35 m and a length of 2.63 m. The floor consists of a 1.87 m diameter endplate that allows the model to be oriented upwind at any angle of incidence when the tests are realized with two rotating cylinders.

The fan-motor consists of a direct current electrical motor of 100 HP and a three-bladed propeller 3 m in diameter.

The convergent has a contraction ratio of 12. The diffuser has an opening angle of 7°. These characteristics lend very good qualities to the flow of the test tunnel.

The critical Reynolds number, measured at the turbulence sphere, is $3.6.10^4$.

The intensity of turbulence in the center of the tunnel is 3.10^{-3} at a flow speed of 40 m/s.

The vertical distribution of the flow velocity is uniform outside the boundary layer, which is around 60 mm thick. To protect against the boundary layer effect, the models are installed on disks placed at a height of 80 mm from the floor. On the disk used, which has a diameter of 800 mm, the thickness of the boundary layer is only 6 mm.

The reference static pressure is measured at the center of the tunnel. The speed of the flow is measured by a Pitot tube placed far in front of the model. The air density is calculated for each series of tests.

A close-meshed net is placed upstream of the test tunnel to obtain a regular flow for wind speeds of less than 5 m/s.

3.2. GLOBAL AERODYNAMIC EFFORTS MEASURES

The experimental set-up makes possible to measure the aerodynamic forces on rotating cylinders of finite aspect ratio from 2 to 6 and of infinite aspect ratio (cylinders fitted between to large discs of 800 mm), with or without rotating endplates, as well as to measure the forces in the case of the interaction between two cylinders of 6 and infinite aspect ratio.

3.2 (a) Experimental setup

Two 380 mm high and 50 mm diameter cylinders are made from 3 mm thick stainless steel. A 100 Watts DC drive motor with separate excitation is placed in each cylinder.

The motor is placed on a shaft on which two ball bearings are fixed. They allow good guidance and perfect rotation of the cylinder around its axis. The main axe of one of the cylinders is finished with an assembly that allows it to be adapted to the aerodynamic balance.

This cylinder is placed in the center of the 800 mm diameter boundary layer trap disc.

This disc rests on an 80 mm high plexiglass shaped wedge. Only the part of the cylinder above the disc is affected by the wind.

Photo 1. Experimental setup.

To reduce the aspect ratio of the cylinder, four 50 mm high wedges are inserted between the 80 mm wedge secured to the floor of the wind tunnel, and the boundary layer trap disc. A greater or lesser height of the cylinder is thus masked.

Circular plates of 100 mm of diameter can be fitted to the base and top of the cylinder.

Infinite ratio is approached by placing the cylinder between 2 "walls". For this, a second 800 mm disc is made in a 10 mm thick stainless-steel plate, the edge is chamfered. The upper part of the cylinder fits into this disc. Again, there is no fluid bypass effect at the ends of the cylinder.

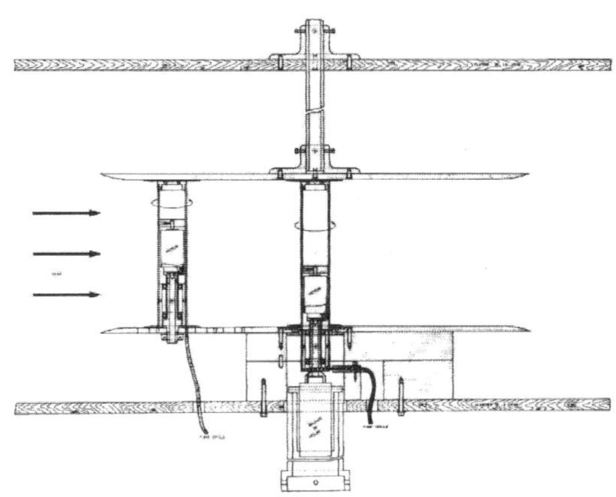

Plan 1: Experimental setup with two cylinders and boundary layer trap disc.

The second cylinder is fixed to the boundary layer trap disc for interaction testing between rotating cylinders. This configuration allows measurements for a geometric aspect ratio of 6 and infinite ratio when the second disc is installed. Both cylinders can be fitted with rotating endplates.

The presence of the lower boundary layer trap disc increases the effective ratio of the cylinder which is close to double the geometric ratio. This experimental configuration makes possible to approach the conditions of use on ships. Flettner rotors will be placed as close as possible to the boat deck to reduce the resistance induced by the geometric ratio.

Photo 2. Experimental setup for two rotating cylinders in infinite aspect ratio.

The reduction in wind speed due to obstruction of the model, a plug effect, is estimated at 10^{-3}.

3.2 (b) Measuring instruments, calibration and operating mode

The cylinder on which the aerodynamic forces are measured is fixed on a balance with gauges having 5 components. The balance measures very small forces, the order of a gram. The output signals from the balance are sent to two SEDEME TS 105 amplifiers and the amplified voltage is read and recorded by a digital multimeter with digital display.

The calibration of the balance consists in measuring the variation of the voltage at the output of the gauge bridges for different loads. Tension independence is verified for several loads placed at different heights along the cylinder and along the two axes, x and y, which will measure lift and drag. The calibration will be repeated several times during the test campaign.

The results are presented as dimensionless coefficients, lift coefficient C_L, and drag coefficient C_D.

The measured drag is the sum of the pressure drag resulting from the forces normal to the cylinder surface, the friction drag which results from the forces tangential to the surface and the induced drag created by the marginal vortexes at the ends of the cylinder.

3.3. PRESSURE MEASUREMENTS

The objective of the experimentation is to study the distribution of pressure on a rotating cylinder as a function of the U/V ratio, as well as the span distribution of the pressure distribution to find the distribution of pressure forces along the span.

The model consists of a cylinder of 130 mm of diameter and 780 mm of height.

The device allows measurements on a rotating cylinder of geometric ratio of 6 and infinite ratio when placed between 2 "walls" of 1 m side. It will be fitted with rotating endplates of twice the diameter of the cylinder.

To avoid the action of centrifugal force on the pressure sensor, this one is fixed on the axis of rotation of the cylinder, and the measurements are carried out continuously.

The sensor is subjected to an input signal which varies over time. The behavior of a sensor in dynamic conditions is determined by a transfer function.

To measure the pressures at different levels of the height of the cylinder, it is sufficient to translate the mast with the help of a rack. The sensor is connected to the wall by a duct of small diameter. This duct rotates with the cylinder. The seal between the duct and the sensor is achieved by a Polstra-type rotating joint. The centrifugal force creates a certain vacuum in the duct during the rotation of the cylinder. The vacuum which then exists at the sensor is compensated by a displacement of the zero of the sensor amplifier.

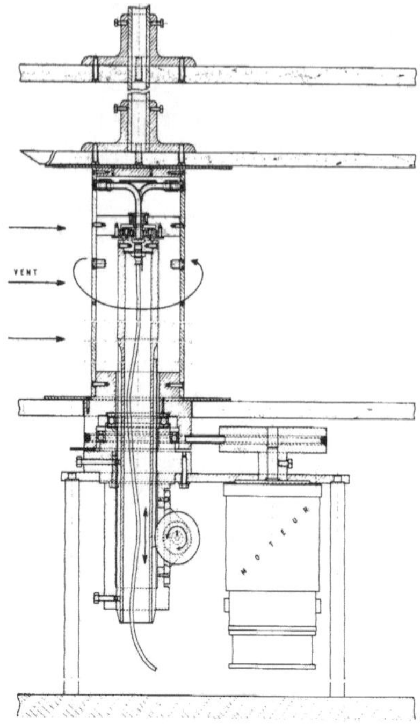

Plan 2: Experimental setup for the measurement of the pressures on the periphery of the cylinder for three heights.

The ENDEVCO 8510-2 sensor selected for the experiment has the following characteristics:

- Measuring range: 0-140 mbar
- Sensitivity: 160 mV / PSI / 10V
- Resonance frequency: 45 Hz
- Sensitivity to acceleration: longitudinal $3.5.10^{-3}$ mB/g and $2.1.10^{-3}$ lateral

The 500 Watts electric motor drives the cylinder at a maximum rotation speed of 70 rpm.

The measurement recording chain makes it possible to record the pressures every 3° leading to 120 measurement points on the circumference of the cylinder.

Several types of input tests are used to characterize the dynamic performance of sensors, they are a step, a Dirac function pulse or a harmonic function.

The most suitable test is the harmonic function because the pressure measured on the rotating cylinder approaches a sinusoidal curve.

The calibration enables the transfer function of the sensor to be determined under the experimental conditions. In particular, to find the frequency response curve of the system consisting of the sensor, the dead volume, the duct connecting the sensor to the wall, and the rotating joint.

The pressure generator designed by ENSAM-Paris is intended for the dynamic calibration of pneumatic lines. It consists of a closed cavity in which there is a sinusoidal pressure created by the movements of the membrane of a speaker.

Several duct diameters are tested. The gain curve of the 5 mm duct has a lower amplitude at resonance than that obtained with the 3 and 4 mm ducts. The resonance peaks appear around 350 and $1.3.10^3$ Hz. Moreover, the gain is practically zero up to 100 Hz. The phase curve of the 5 mm duct shows that the phase shift only appears from a frequency of 100 Hz.

The level of the fundamental, fixed by the speed of rotation of the cylinder that does not exceed 70 rpm (70 Hz), will not be disturbed by the experimental setup. To ensure the accuracy of the results, especially for the phase shift, the Fourrier series development of the pressure distribution will be pushed to 30 harmonics.

The pressure measurements are related to the static pressure Po.

The pressure coefficient curves Cp are plotted as a function of the angle θ:

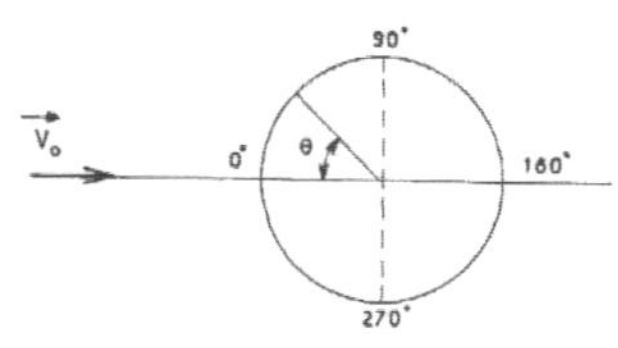

Figure 1 gives the distribution of the pressure on a rotating cylinder corrected for the transfer function. The cylinder of ratio 6 has two rotating endplates, it rotates at 18 rpm, Re = $9.17.10^3$ and U/V = 0.7. The pressure tap is halfway up the cylinder. The amplitude fluctuations are smoothed and the phase shift corrected.

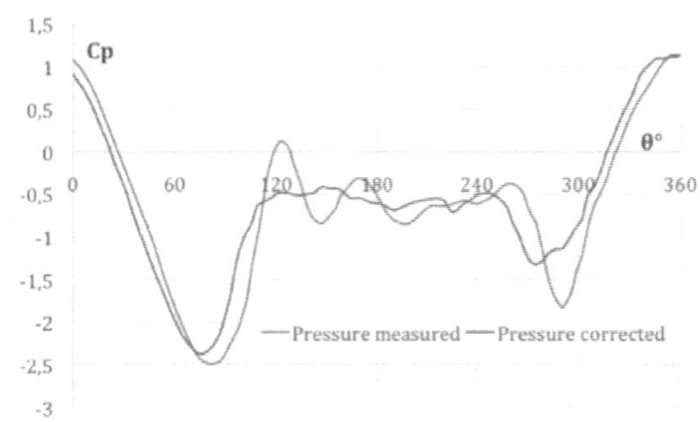

Figure 1: Pressure measurements on the periphery of the cylinder without and with correction of the signal by the transfer function. λ=6, U/V = 0.7, Re = $9.17.10^3$.

4. RESULTS

4.1. MEASUREMENTS OF GLOBAL AERODYNAMIC EFFORTS ON A ROTATING CYLINDER

The purpose of these measurements is to demonstrate the effect of the aspect ratio, the effect of the rotating endplates and the influence of the Reynolds number.

The aerodynamic forces are presented in the form of polar $C = f(C_D)$, lift coefficient as a function of drag coefficient, and graduated in k = U/V, the ratio of the peripheral speed to the speed of the wind in the wind tunnel.

The literature shows that the aerodynamic performance of rotating cylinders depends on the configurations selected: aspect ratio, rotating endplates or not, values of the U/V ratio and of the Reynolds number.

Figure 2 gives the polars obtained with cylinders of aspect ratio of 3 to 6 and in infinite aspect ratio, with rotating endplates. The Reynolds number of the tests is $1.35.10^3$.

The lift coefficient is not very sensitive to the aspect ratio of the cylinder from infinity to 3. The drag coefficient is multiplied by 6 between the infinite aspect ratio to 3. The maximum C_L / C_D ratio goes from 10 for the infinite aspect ratio to 4 for the aspect ratio of 6. Lift coefficients of 12 are measured.

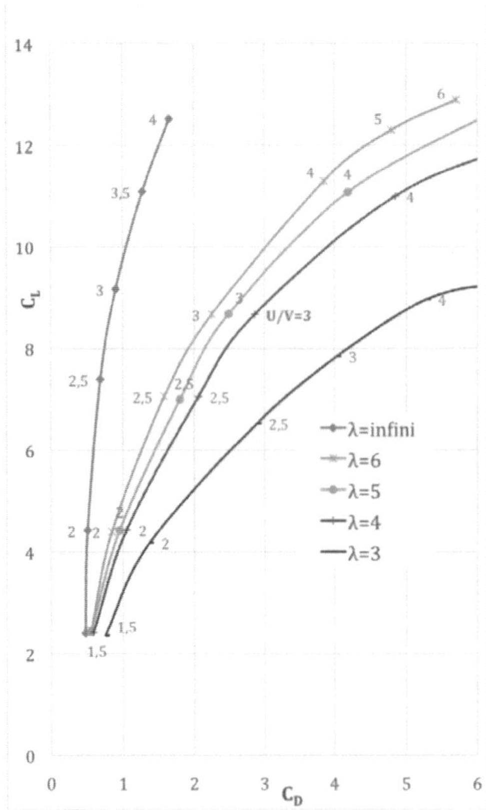

Figure 2: Polars obtained on rotating cylinders fitted with endplates of infinite aspect ratio and aspect ratio of 3, 4, 5, and 6.

4.1 (a) Influence of the Reynolds number

For the wind propulsion of ships, systems with high lift and the lowest possible drag are sought (optimization of performance at up-windward), only the results for aspect ratios greater than 5 are presented in the publication and for U/V greater than 1.5 (beyond the inversion of the Magnus effect).

The test configurations found in the literature are highly variable and comparisons of results difficult to interpret. In figure 3, the values of C_D obtained by the measurement of the global forces by Bordogna (Bordogna, 2019) and Charrier (Charrier, 1979) for very similar configurations are presented. For Bordogna, geometric ratio $\lambda = 4$ between walls (simulation infinite ratio), without endplate, and for Charrier (Charrier, 1979), geometric ratio of 6 between large boundary trap discs without endplate, only the Reynolds numbers are different $1.35.10^3$ for Charrier and more than 3.10^5 for Bordogna. The relatively small differences between C_D in the $2 < k < 4$ operating range show that the tests performed at low Reynolds are relevant.

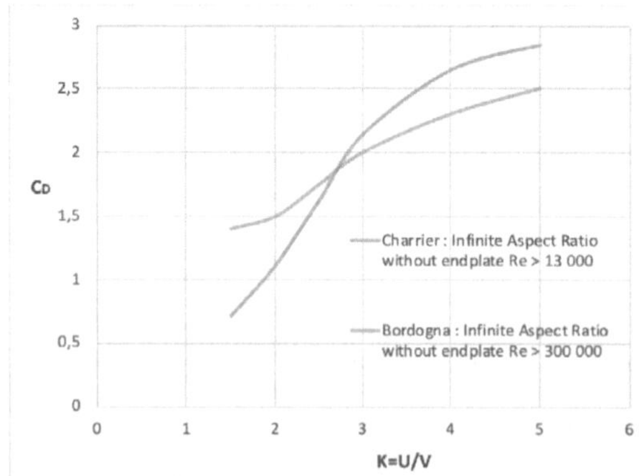

Figure 3: Comparison of C_D as a function of U/V measured in infinite ratio without endplate by Charrier (Charrier, 1979) and Bordogna (Bordogna, 2019) for two values of the Reynolds number.

4.1 (b) Influence of endplates in infinite ratio

Figure 4 shows the polars $C_L = f(C_D)$ obtained by Bordogna and Charrier in infinite ratio and shows the important role of the rotating endplates. The values of C_L are practically independent of the Reynolds number. It is a classic result obtained for airplane wing profiles.

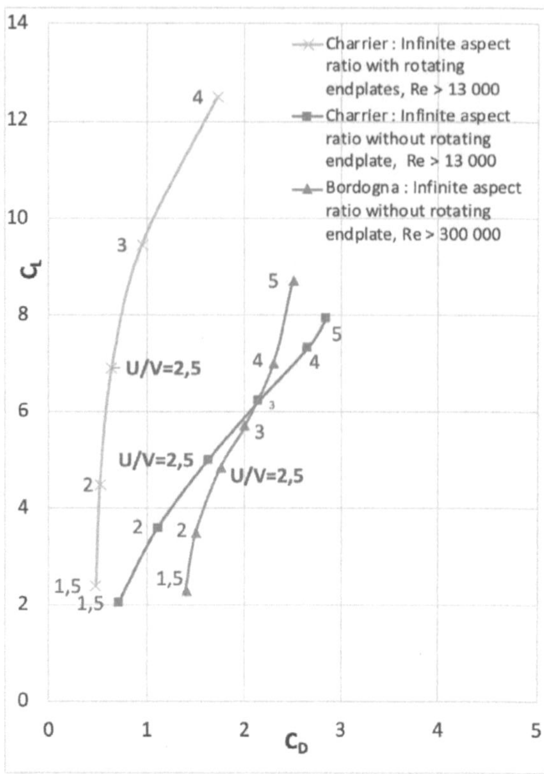

Figure 4: Polars $C_L = f(C_D)$ obtained in infinite aspect ratio without and with rotating endplates by Charrier and Bordogna.

4.1 (c) Influence of the endplates for a aspect ratio of 5

To get closer to commercial conditions of use, tests were carried out with a rotating cylinder with geometric ratio of 5. The bottom of the cylinder is in the center of the boundary layer trap disc, which has the effect of increasing the effective aspect ratio of the cylinder which should approach 10.

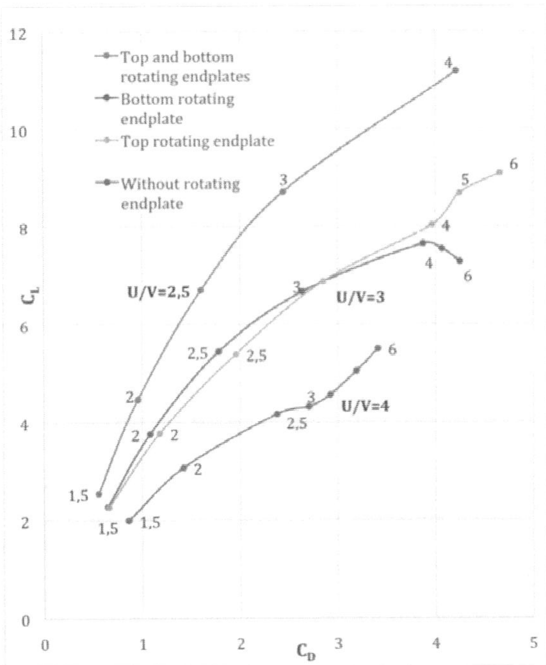

Figure 5: Comparison of the polars $C_L = f(C_D)$ obtained in aspect ratio 5 with and without endplate.

As shown in Figure 5, in finite ratio of 5, the C_L with endplates can be double of the C_L obtained without endplate for the same value of k between 2 and 4.

Without endplate, the C_D is about 50% higher than with endplate.

The results show that the rotating endplates, top and bottom of the cylinder have an identical influence on the C_L and the C_D. The flow at the ends is very strongly modified by the endplate far beyond the usual effect of the winglets that decrease the drag induced by the lift by limiting the intensity of the vortex at the extremities of the wings. The rotating endplates play an important role of dragging the fluid by friction resulting in an increase in fluid velocity and a rolling effect of the fluid on the cylinder. This phenomenon will be clarified when measuring the pressure along the span of the cylinder for $\lambda = 6$.

4.2. PRESSURE MEASUREMENTS ON A ROTATING CYLINDER

This publication presents the results obtained on a cylinder of infinite ratio and geometric aspect ratio of 6, equipped with rotating endplates with a diameter double of the diameter of the cylinder. The bottom of the cylinder is placed in the center of a boundary layer trap wall and a second wall is placed on the top of the cylinder to simulate infinite ratio. The two walls are one meter of sides.

The pressure measurements are made in three positions: in the middle of the cylinder (position 5), three quarters of the height (position 3), and 2 cm from the upper endplate (position 1).

Figure 6 gives the distribution of the pressures in the middle of the rotating cylinder (position 5) for values of k = U/V from 2.5 to 5. The peak of depression is obtained at azimuth 100 ° - 105 °. The "dead" fluid zone, Cp = 0, appears at 180 ° when U/V = 2.5. This zone starts further when U/V increases, i.e. 220 ° for U/V = 5. When U/V exceeds 2.5, the air streams are forced to follow the surface of the rotating cylinder well beyond 180 °, dramatically increasing lift. For these tests, the Reynolds number is between $4.75.10^4$ and $7.1.10^4$.

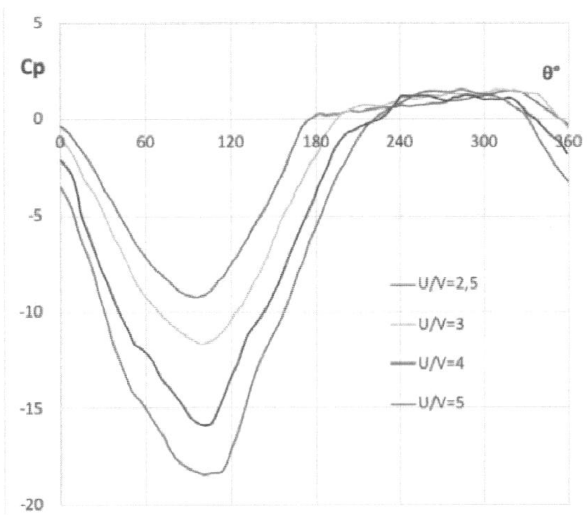

Figure 6: Distribution of the pressures measured in the middle (position 5) of the rotating cylinder for 4 values of U/V.

Figure 7 gives the polars obtained by integration of the pressures measured in position 5, in the middle of the cylinder and by measuring the overall forces with the aerodynamic balance, for rotating cylinders in infinite ratio and of aspect ratio of 6. The polars are not superimposed nor for infinite ratio or aspect ratio of 6. If the lift coefficient, evaluated by integration of pressures or measured directly through the balance, get a value, more or less the same for each k, it is not the case for the drag coefficient. The drag coefficient measured by the balance is largely bigger than those measured with the balance. This is observed for the infinite ratio as well as for the aspect ratio of 6. The integration of the pressures at the mid-height of the cylinder does not allow to evaluate the real effort along the entire span of the rotating cylinder.

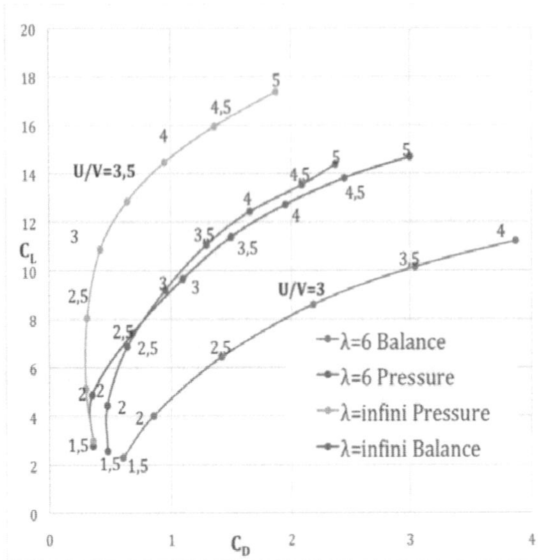

Figure 7: Comparison of the polars $C_L = f(C_D)$ obtained by integration of the pressures and with the aerodynamic balance for two different configurations.

Figure 8 shows the pressure coefficient (Cp) measured at positions 1, 3 and 5 for U/V = 4. The peak of vacuum shifts from 100 ° to 120 ° as the pressure tap approaches the upper endplate. Everything happens as if the fluid was spinning. The fluid twins around the cylinder driven by the endplates and the twisting is more and more important close to the ends. The phenomenon must also occur on the lower part of the cylinder. Only the central part of the cylinder is not influenced by the endplates. The same winding phenomenon is observed for all U/V values between 1.5 and 5. The results are not given is this publication.

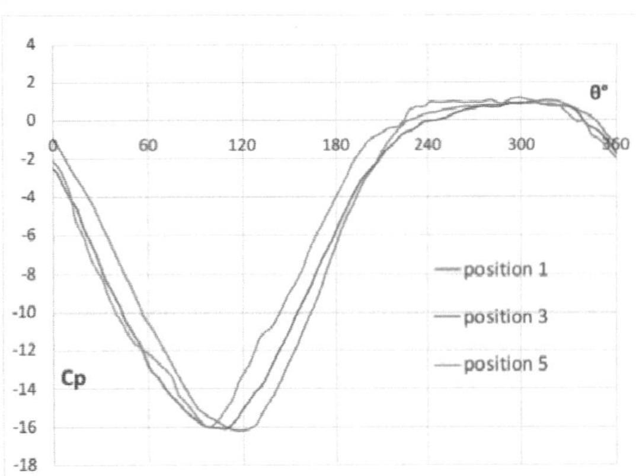

Figure 8: Pressure measurements in three positions of the span of the rotating cylinder for U/V=4.

The lift coefficient obtained by integration of the pressures, remains constant for the three positions 1, 3, and 5 for the U/V values of 1.5 to 5.

Figure 9 compares the lift coefficient obtained by integration of the pressures for position 5 with that obtained by the measurement with the aerodynamic balance of the overall forces.

The difference in C_L obtained by the two measurement means is less than 10%.

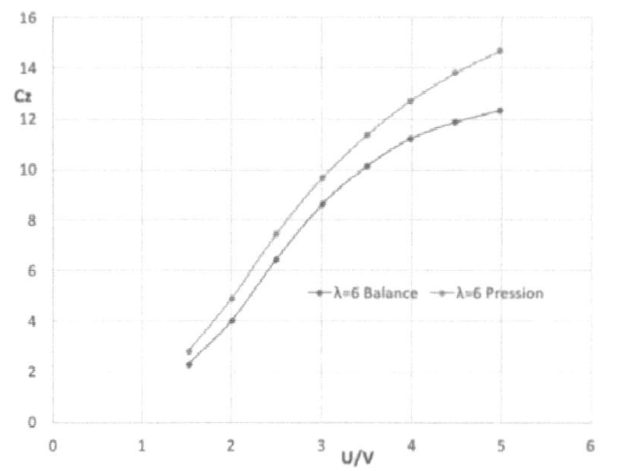

Figure 9: Comparison of the lift coefficient obtained by integration of the pressures for position 5 and by the aerodynamic balance.

The results are very different with regard to the drag coefficient C_D.

Figure 10 gives the C_D obtained by integration of the pressures for the three positions 1, 3 and 5 compared to those obtained with the aerodynamic balance. The C_D increases steadily from the central position to the upper end of the rotating cylinder, regardless of the value of U/V from 1.5 to 5.

This confirms that the fluid is winding in span and that the top of the cylinder is attacked by a lower incidence wind than its middle. The same phenomenon should occur in the lower part of the cylinder that is also equipped with a rotating endplate.

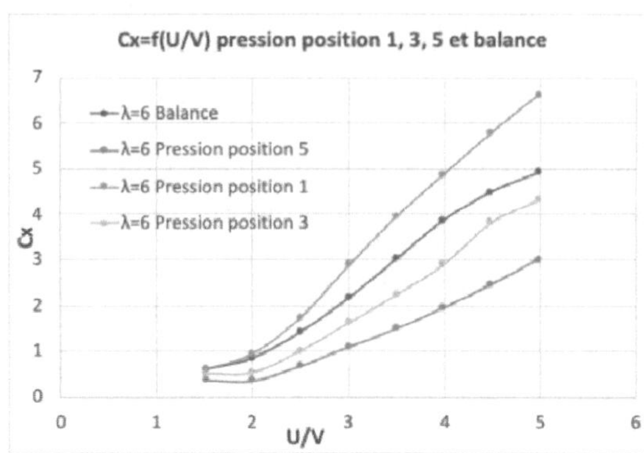

Fig 10. Influence of the winding on the drag coefficient measured at three positions of the cylinder span.

These observations and measurements lead to the following affirmations that:

• The integration of the pressures according to the circumference of the cylinder does not make possible to obtain exact values of the lift and drag coefficients of the rotating cylinder whether or not it is equipped with rotating endplates.

• The rotating endplates have a considerable influence on the aerodynamic performance of the rotating cylinder, far beyond the reduction of induced drag by the aspect ratio effect.

• The endplates drive the fluid by viscosity in a spiral motion, winding around the rotating cylinder.

• The experimental approach on models representing the real configurations as faithfully as possible, must be favored to understand the aerodynamic performance of the wind propeller by rotating cylinders.

4.3. MEASURES OF THE INTERACTION BETWEEN TWO ROTATING CYLINDERS

In many practical applications, two or more cylinders will be combined. It is therefore important to quantify the interaction between them. This study evaluates the interactions between two cylinders of infinite aspect ratio and aspect ratio of 6 equipped with rotating endplates.

In figure 11 are plotted the polars $C_L=f(C_D)$ obtained in the case of the interaction between two cylinders of aspect ratio of 6 equipped with two rotating endplates having clockwise rotation, as a function of β, angle formed by the line between the two cylinders and the wind direction. The forces are measured on cylinder 2.

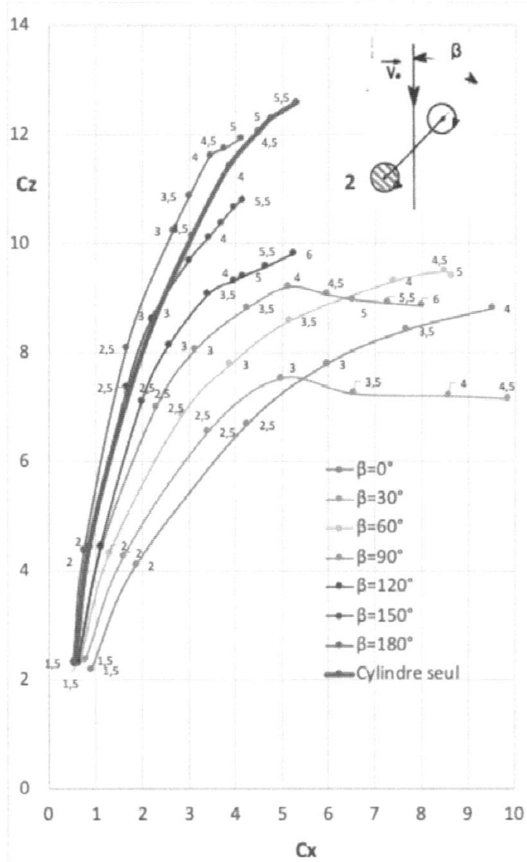

Fig 11. Polars obtained on cylinder 2 for 0 <β <180 ° in interaction with the other rotating cylinder.

Whatever the values of U/V above 1.5 and whatever the value of the angle β, the lift coefficient is always significantly lower than that of the cylinder alone. The C_L of cylinder 2 finds the value of C_L of the cylinder alone for β = 150 °.

The coefficient C_D is always greater than that of the cylinder alone and can reach more than double for β smaller than 60 °. This is very detrimental for up-windward, when minimal drag of the wind propeller is sought.

The forces on cylinder 1 are only measured for a few values of β. For β greater than 30 °, the C_L is greater than that of the cylinder alone, whatever U/V. The C_D is slightly higher than that of the cylinder alone.

Figure 12 gives the results for β = 90 °.

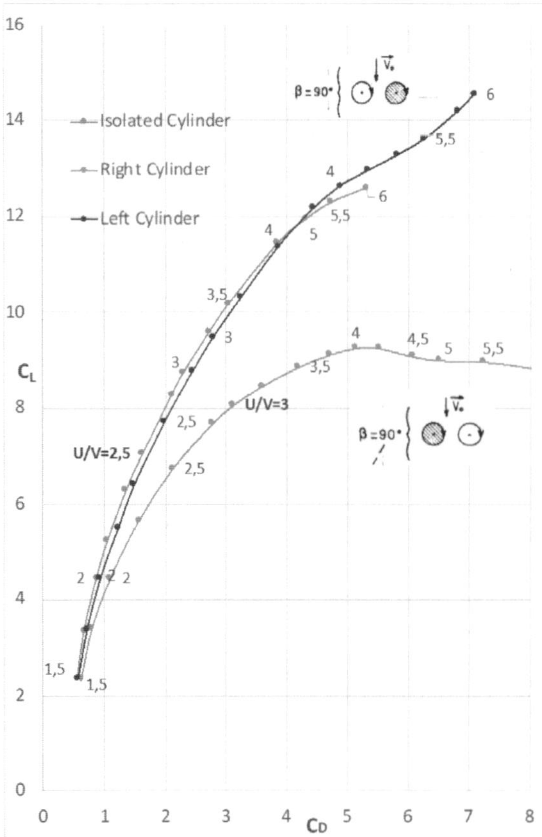

Fig 12. Comparison of the polars obtained on the two cylinders perpendicular to the wind

Photo 3 shows the visualization of the flow on cylinder 2 when β = 90 ° and U/V = 4.5, and photo 4 when the cylinder is alone for the same U/V = 4.5. The flow comes from the left and the cylinders rotate clockwise. This is the configuration of Figure 12.

The effect induced by cylinder 1 is a decrease in the intensity of the wind velocity hitting cylinder 2, and a strong inclination of the flow behind the cylinder.

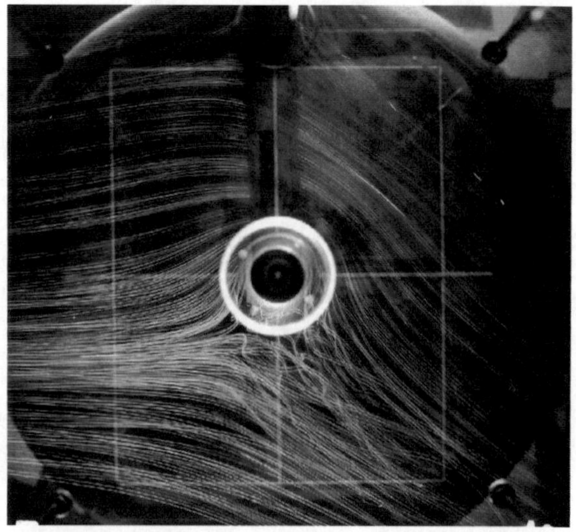

Photo 3. Visualization of the flow on cylinder 2 (middle of the picture) interacting with cylinder 1 (top of the picture) for U/V=4.5.

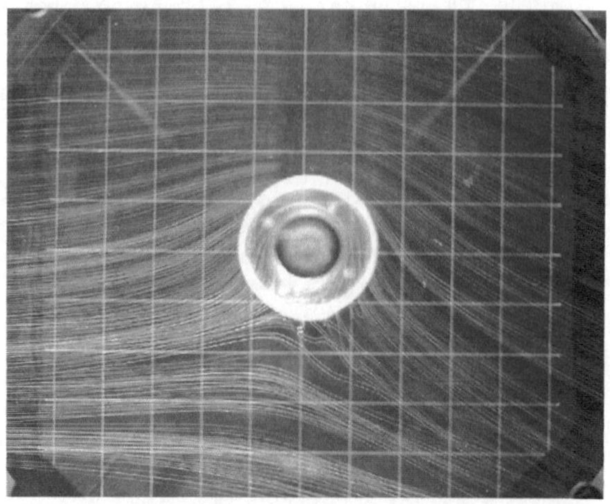

Photo 4. Visualization of the flow on an isolated cylinder for U/V=4.5.

In summary, the effect induced by cylinder 1 on cylinder 2 is a decrease in C_L and a large increase in C_D, but the effect of cylinder 2 on cylinder 1 is a slight increase in C_L and a significant increase in C_D.

Overall, the interaction between two cylinders of aspect ratio of 6, having clockwise rotation and at the same speed, equipped with endplates, is a conservation of lift and a strong increase in drag compared to the cylinder alone.

In applications where minimal drag is sought, such as for the propulsion of ships, the cylinders should be separated by a minimum distance of at least 10 times the diameter of the cylinders.

5. CONCLUSION

The aerodynamics of thick profiles whose relative thickness is greater than 50% as for the Suction Sail or equal to 100% for the Flettner Rotor, calls upon physical mechanisms very different from those of thin profiles, usually used in aeronautics or for sailing boats.

The energy expended for the rotation of the cylinder or that used to control the boundary layer detachment for the Suction Wing, dramatically changes the aerodynamic theories conventionally used to predict airfoil and wing performance. In the case of the rotating cylinder, the role of viscosity is preponderant.

In finite aspect ratio, at the ends of the rotating cylinder, the pressure suddenly passes from very high negative value on one side of the system to positive value or zero on the other side causing the creation of vortex structures of very high intensities. The presence of the rotating endplates limits the intensity of the end vortex structures and plays a very important role in increasing lift and maintaining the steady distribution of lift along the span of the cylinder. But with that comes a huge increase in drag at the ends of the rotating cylinder.

The fluid wraps around the cylinder, it is driven in a spiral modifying the angle of incidence of the wind on the cylinder over more than a third of the height at both the top and bottom of the cylinder.

These considerations suggest that measuring the pressures on a rotating cylinder with or without endplate is not sufficient to assess the aerodynamic performances of the cylinder in infinite or finite aspect ratio.

The interaction between two cylinders with an aspect ratio of 6, rotating clockwise at the same speed, spaced by 6 diameters depends on the value of β, angle formed by the line between the two cylinders and the wind direction. Regardless of the angle β, the overall lift of the two cylinders is almost twice the lift of an isolated cylinder. The overall drag is very strongly increased, more than 100% for β between 30 and 60 °.

To reduce the interaction between the rotating cylinders, they should be spaced at least 10 diameters.

Knowledge of the aerodynamic characteristics of the isolated cylinder is not sufficient to estimate the performance of a system composed of several rotating cylinders. Only wind tunnel tests in real configurations of the use of the Flettner rotor wind propeller will provide better accuracy of its propulsive power.

6. ACKNOWLEDGEMENTS

I would like to thank former and new actors of maritime transportation sector who are convinced that the use of wind energy adapted to ships could significantly contribute to the reduction of greenhouse gas emissions and directly limits the global warming, one of the biggest threat on the habitability of the planet.

7. REFERECES

[1] Von G. Magnus, (1853), *Ueber des Entstchen von theer aus olbidendem Gase*, Annalen der Physik und Chemic n°1.

[2] Lord Rayleigh, (1857), *On the Irregular Flight of a Tennis Ball,* Scientific paper, vol.1.

[3] Lord Rayleigh, (1877), Messenger of Mathematics, vol.7.

[4] A. Thom, (1926) *The Aerodynamics of the Rotating Cylinder,* Thesis, University of Glasgow.

[5] A. Thom, (1925), *Experiments of the Air Forces on Rotating cylinders*, Aeronautical Research Committee, P and M1018.

[6] R.W.Van Aken and H.R. Kelly, (1957), *The Magnus Effect on Spinning Cylinders,* Reprint n°712, 25th Annual Meeting T.A.S..

[7] S.W.Swanson, (1956), *An Experimental Investigation of the Magnus Effect,* Final Report, 00R Project n°1082, Case Inst. of Tech..

[8] Bertrand Charrier, (1979), *Etude Théorique et Expérimentale de l'Effet Magnus, Destinée à la Propulsion Eolienne de Navires*, Thèse de Physique, Université de Paris VI, Paris.

[9] Von A. Flettner, (1925), *The Flettner Rotor Ship*, Engineering, vol.19.

[10] B. Charrier *et al. (1985), Foundation Cousteau and Windship Propulsion,* Journal of Wind Engineering and Industrial Aerodynamics 20.

[11] Lucien Malavard, (1984), *Un nouveau Propulseur Eolien de Navires*, Comptes Rendus de l'Académie des Sciences, Tome 1, n°1.

[12] C. Badalamenti, and S. A. Prince, (2008), *The Effects of Endplates on a Rotating Cylinder in Crowflow*, 26th AIAA Applied Aerodynamics Conference, Honolulu.

[13] G. Bordogna, *et al*., (2019), *Experiments on a Flettner Rotor at a Critical and Supercritical Reynolds Numbers*, Journal of Wind Engineering and industrial Aerodynamics 188.

[14] A. De Marco, et al., (2016), *Flettner Rotor Concept for Marine Applications: A Systematic Study*, International Journal of rotating Machinery, Article 3458750.

[15] V. Paakkari, A. Hurford, C. Craddock, (2019), *Rotor Sail, GHG Reduction Potential, Modelling and Sea Trial Validation*, Wind Propulsion Conference, The Royal Institution of Naval Architect.

FLETTNER ROTORS PERFORMANCE AND INTERACTION EFFECTS ON THE MARIN HYBRID TRANSITION COASTER

M Garenaux and J J A Schot, MARIN, The Netherlands

SUMMARY

In the context of ship design, including wind propulsion, knowing the aerodynamic performance of a ship and its wind propulsors is essential. In the literature, the performance of wind propulsors is generally derived from wind tunnel test data, full scale tests or CFD. Those experimental or numerical databases are often omitting the interactions with the ship and its superstructures and atmospheric boundary layer. In this research, a numerical approach using RANS simulations is developed to estimate the effects of the interaction between the MARIN Hybrid Transition Coaster (MHTC) and its three Flettner rotors. In order to assess the effects of these interactions, the performance of a Flettner rotor positioned on the deck of the vessel is compared to the performance of a free standing Flettner rotor. Large interaction effects are found to depend on the apparent wind angle. Accounting for these interactions is essential. A model is proposed to correct the performance of a free standing Flettner rotor to the performance of a rotor on the deck of the vessel. In addition the effects of multiple wind propulsors are studied in relation to this model.

NOMENCLATURE

Vs	Ship speed (kn)
TWS	True wind speed (m/s)
TWA	True wind angle (deg)
AWS	Apparent wind speed (m/s)
AWA	Apparent wind angle (deg)
β	Drift angle (deg)
Ψ	Heading (deg)
n	Hellman exponent (-)
f_{ws}	Factor on wind speed due to interaction (-)
Δ_{wa}	Offset in wind angle due to interaction (deg)
MHTC	MARIN Hybrid Transition Coaster
RANS	Reynolds Average Navier-Stokes equations
CFD	Computational Fluid Dynamics
EARSM	Explicit Algebraic Reynolds Stress Model
ABL	Atmospheric Boundary Layer
$N_{flettner}$	Flettner rotation rate (RPM)
SR	Spin Ratio
D	Rotor diameter
d	Gap distance between two rotors
PPP	Performance Prediction Program
C_L	Lift coefficient
C_D	Drag coefficient
C_T	Total force coefficient
C_X	Driving force coeffcient

1. INTRODUCTION

This paper presents the wind propulsor interaction effects on the MARIN Hybrid Transition Coaster (MHTC). The interaction effects are determined using RANS calculations and used to validate the assumptions made in a simplified modelling framework. The proposed modelling framework is described in Paragraph 2 and is required because interaction effects often have to be incorporated in wind propulsion performance predictions in which calculations have to be performed for various combinations true wind angle and wind speed. For each of these wind conditions the optimum sailing equilibrium has to be solved which requires fast numerical methods. The used MHTC test case is described in Paragraph 3. The accuracy of the RANS calculations is determined using a verification study for the numerical setup and validation of the outcome is performed using experiments for the rotor-rotor interaction at model scale. This is described in Paragraph 4. The results of the calculations are used to validate the assumptions in the modelling framework and to assess the effect of the interaction on the driving force for the MHTC test case in Paragraph 5.

2. THEORY

The basis for the proposed modelling framework is the assumption that the change in forces produced by the wind propulsion device is caused by a change to the apparent wind speed (AWS) and apparent wind angle (AWA) due to the presence of the ship hull, superstructure and other wind propulsors. This results in the effective wind speed and angle. This is then used to determine the wind propulsor forces including interaction using the performance curve of a standalone unit in an uniform wind profile. In general these interaction effects are assumed to be build up from four contributions:

1. Effect of the atmospheric boundary layer wind profile
2. Effect of the ship speed on the true wind resulting in the apparent wind and twist
3. Effect of the superstructure on the apparent wind resulting in the disturbed wind
4. Effect of the wind-propulsors on each other resulting in the induced wind

The first three effects are expected to have only a one-way interaction with the wind-propulsor devices. Meaning that the local wind is affected, but the effect of the wind propulsor back on this local wind is neglected. The last item depends also on the lift produced by the wind propulsors and therefore on the operational setting of each wind propulsor such as the rotation rate of rotors. Furthermore, the approach assumes that the interaction can be described by the change in wind at a single point. For wind propulsors that depend strongly on the angle of attack, this assumption should be validated as the twist in the apparent wind profile might be more important in that case.

2.1 Effect of the atmospheric boundary layer wind profile

The true wind speed is defined here at a height of 10 meters above the water surface. The wind velocity is assumed to follow an atmospheric boundary layer profile determined by a power coefficient as shown in Equation (1). This power coefficient is called the Hellmann exponent and often set to 1/7 [1],[2]. The result is a varying wind speed over the height of the wind propulsor depending on the location of the base of the device from the water surface. As the largest part of the device is often above 10m the effective wind speed experienced by the device is generally increased due to this effect.

$$TWS_z = TWS_{z=10m}\left(\frac{z}{10.0}\right)^n \tag{1}$$

2.2 Effect of the ship speed on the true wind resulting in the apparent wind and twist

The wind experienced by the wind propulsors is a summation of the true wind speed and the wind resulting from the ship speed. The combined wind is called the apparent wind. The true wind speed varies over the height as a result of the atmospheric boundary layer while the wind generated by the ship speed is uniform. This leads to non-uniform wind angles over the height, called twist. The used definitions for the wind angle are shown in Figure 1.

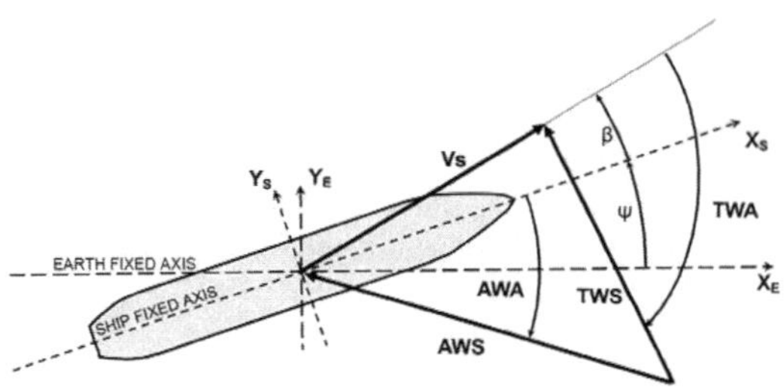

Figure 1: Used definitions for the wind angles.

This interaction with the forward ship speed complicates the modelling. Depending on the vertical location of the wind propulsor the apparent wind angle will vary. The apparent wind angle is however used as input variable to describe the interaction with the superstructure and between the wind propulsors, which is described later on in this paragraph.

In order to have a consistent reference for the apparent wind angle it is also defined at a height of 10m. The effect of the change in height of the wind propulsor on the apparent wind angle and speed is described by a factor on the wind speed and an offset in wind angle as shown in Equation (2) and (3). It should be noted that it is here assumed that it is sufficient to describe the interaction at a single point. If the effect of twist is significant the interaction coefficient will not be constant with varying ship speed or Hellmann coefficient. Later research should show if the interaction is independent also for devices more sensitive to changes in the angle of attack than Flettner rotors.

$$f_{ws\,z} = \frac{AWS_z}{AWS_{z=10m}} \qquad \Delta_{wa\,z} = AWA_z - AWA_{z=10m} \tag{2}$$

$$AWS_z = f_{ws\,z}\,AWS_{z=10m} \qquad AWA_z = \Delta_{wa\,z} + AWA_{z=10m} \tag{3}$$

2.3 Effect of the superstructure on the apparent wind resulting in the disturbed wind

The apparent wind is disturbed by the superstructure of the vessel before reaching the wind propulsor. This effect is assumed to depend on the position of the wind propulsor on the vessel and the apparent wind angle. It is assumed that the effect of operating setting of the wind propulsor and twist can be neglected, using only the apparent wind angle at 10m. The interaction effect can be determined by taking the difference in forces between the vessel with and without the wind propulsor.

As shown later, the atmospheric boundary layer is important for accurately calculating the flow around the hull and superstructure [3]. The atmospheric boundary layer should therefore be included in the calculation. As the effect of the atmospheric boundary layer is already included in the previously described components of the interaction coefficients its effect should be removed from the results as shown in Equation (4).

$$f_{ws\,disturbed} = \frac{f_{ws}}{f_{ws\,z}} \qquad \Delta_{wa\,disturbed} = \Delta_{wa} - \Delta_{wa\,z} \tag{4}$$

2.4 Effect of the wind-propulsors on each other resulting in the induced wind

The effect of wind propulsors on each other is assumed to be primarily a result of the induced velocities, although the wind shadow could also have some effect. This effect also depends on the operational settings determining the lift of the device. This increases the amount of conditions that should be calculated to describe the interaction coefficients of this aspect. In practical applications it might therefore be required to use a fast lower fidelity method such as the lifting line approach to obtain these coefficients.

The combined interaction coefficients are finally described by Equation (5) assuming that the contributions are independent and can be combined.

$$f_{eff} = f_{ws\,z}\, f_{ws\,disturbed}\, f_{ws\,induced} \qquad (5)$$
$$\Delta_{wa\,eff} = \Delta_{wa\,z} + \Delta_{wa\,disturbed} + \Delta_{wa\,induced}$$

3. MHTC test case

The MARIN Hybrid Transition Coaster (MHTC) concept design is used as test case to investigate the effects of hull and superstructure and wind propulsor interaction on the wind propulsion performance. The dimensions of the full scale Flettner rotors used on the vessel are given in Table 1. The rotor is mounted on a planar surface (simplified deck of the vessel). The design represents a coaster with a length between perpendiculars of 86.76 m, a beam of 13.99 m and a freeboard of 7.0 m with a simplified superstructure. The vessel is fitted with three rotors which are separated by 28.27m and the aft rotor is positioned 26.69m ahead of the aft perpendicular. The gap distance, defined centre to centre, is roughly d/D≈9.5. Based on previous literature [4] limited rotor-rotor interaction is expected. An illustration of the complete configuration is given in Figure 2.

Table 1: Dimensions of full scale Flettner rotor

Quantity	Symbol	Value	Unit
Rotor height	H_{rotor}	18.0	[m]
Rotor diameter	D_{rotor}	3.0	[m]
Disc thickness	t_{disc}	0.1	[m]
Disc diameter	D_{disc}	5.5	[m]
Lateral area	A_{rotor}	54.0	[m2]
Aspect ratio	AR	6.0	[-]

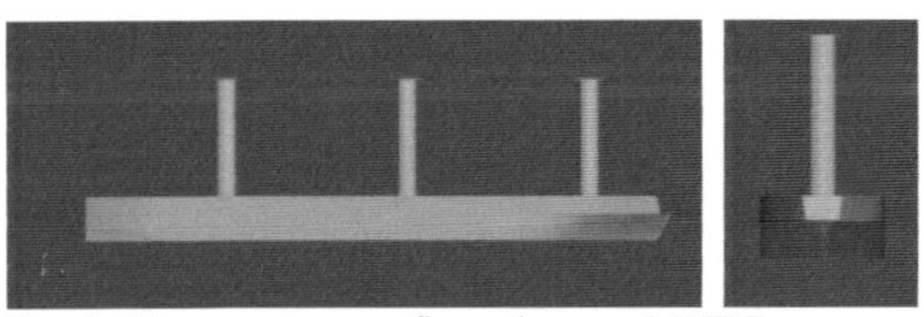

Fig. 2: Flettner rotor configurations on MHTC.

4. Verification and validation of numerical setup

The interaction effects are studied for the MHTC using steady full scale RANS. A verification study is performed for the numerical setup used for the single rotor and a single rotor on the deck of the MHTC. Furthermore, validation is performed using model scale experimental data from [4]. Both experimental data for a standalone Flettner rotor and rotor-rotor interaction have been used for this comparison.

4.1 Computational domain and grid

For the region close to the rotors a cylindrical grid topology is used, because it was observed that this leads to a significantly smaller numerical error compared to a cartesian grid of similar grid size. Further away a cartesian grid is used and coupled using interfaces with second order interpolation. For the standalone rotor a large rectangular domain was used to avoid effects of the sides of the domain. The dimensions are 300 rotor diameters long, 200 wide and 100 high. For the simulations with MHTC superstructure, a circular background domain around the vessel is used. The diameter of the domain is 170 rotor diameters. The near wall region is resolved near the rotor (y+ <1). The used discretisation is visualized in Figure 3.

Fig. 3: Discretisation for the single rotor (top) and rotors on the MHTC (bottom) showing the regions with a cylindrical grid in red.

4.2 Numerical settings

A no slip wall boundary condition is used on the superstructure and a no slip rotating wall condition on the rotor. For the simulation with standalone rotor, a velocity inlet boundary condition is used at the front of the domain. A pressure boundary condition is used at the sides, outlet and top of the domain. At the bottom of the domain two boundary conditions are used: a free slip wall is used for the background part of the domain and a no slip wall boundary condition near the foot of the rotor. For the simulations with the MHTC superstructure, an inflow boundary condition is used on the side of the circular domain allowing to compute wind conditions from all directions. At the bottom of the domain, a no slip wall is defined.

Unless otherwise specified all simulations, model and full scale, are conducted with the Explicit Algebraic Reynolds Stress Model (EARSM). It must be acknowledged that recent findings suggest that the kω-SST turbulence model yields better results at full scale. For this reason the rotor-rotor full scale interaction as well as the support structure full scale interaction are determined using the kω-SST turbulence model.

An atmospheric boundary layer (ABL) profile is specified at the inflow boundary conditions of the simulations. In addition to the speed profile a free stream turbulence condition is specified based on the work described in [5]. It is mandatory that the turbulence values associated with the ABL profile are not used in the calculation. This is discussed in details in [5]. The length scale of the turbulent eddies associated with the ABL are very large compared to the ship and rotor and would not interact with the small boundary layer forming on the vessel and rotor. If included in the RANS simulations these turbulent values would cause artificially high turbulent viscosity instead of large scale unsteady inflow velocities.

4.3 Grid sensitivity study

A grid sensitivity study has been carried out for the standalone rotor at model and full scale Reynolds number as well as for a single rotor on the MHTC hull and superstructure at full scale Reynolds number. The results have been presented in [3] and indicated a good convergence behaviour with successive grid refinements. The intermediate grid for the standalone rotor (2,631,176 cells) is used in this study. For the single rotor on the MHTC the second finest grid (18,814,696 cells) is selected for the rest of this study.

4.4 Validation with experimental data

The results of the described numerical approach have been validated against experimental data from the wind tunnel tests performed by Peutz BV as described in [3]. It was concluded that the CFD results correlates well with the experimental data at model scale.

More recently a comparison with experimental data for rotor-rotor interaction from [4] has been conducted. This rotor has a reduced aspect ratio, from 6 to 5, and a slightly larger end plate diameter, of twice the rotor diameter, compared to the MHTC rotor. The geometry of the wind tunnel test section of Politecnico Di Milano is used to define a numerical domain. No slip wall boundary conditions are defined on the wind tunnel walls, floor and ceiling. An inlet boundary condition is specified at the inlet of the wind tunnel. The outlet, numerically positioned 600 rotor diameters downstream, is defined as a pressure boundary condition. Similar to the experiments the rotor(s) is (are) positioned 0.3m above the floor. The support structure(s) of the rotor is (are) simplified in the simulations, using the stationary endplate at the bottom of the rotor.

The comparison with the results of the single rotor used in these experiments indicate a good correlation with CFD. This is presented in Figure 4. A good agreement in both lift and drag coefficients is found between experiments and CFD. Up to SR=3 the differences between both turbulence models are small. Above SR=3 some differences are observed. It must be acknowledged that above SR≈3 to 3.5 a steady state approach theoretically no longer suffices due to vortex shedding and one should switch to an unsteady approach. This has been reported by [6] and concluded from previous internal study at MARIN. For the range of Reynolds numbers computed, the results only show sensitivity to the Reynolds number for spin ratios below 1.25. However, it is known from [4] that the sensitivity is larger for a wider range of Reynolds numbers.

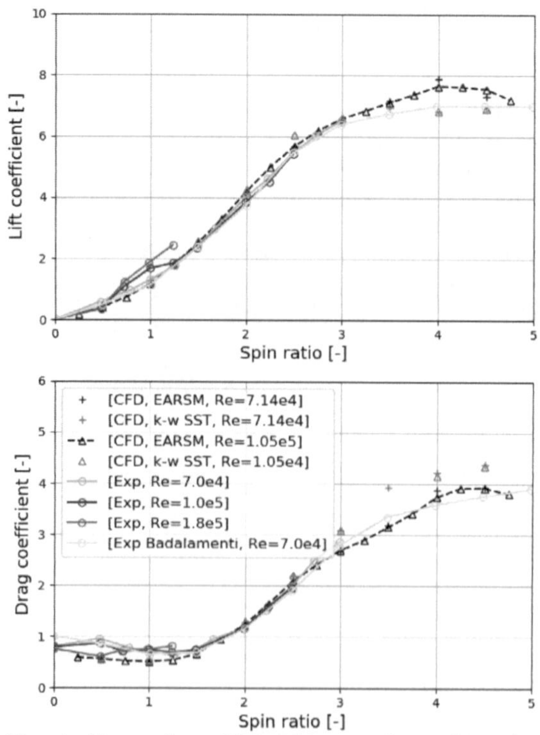

Fig. 4: Comparison CFD with experimental results from [4] .

The rotor-rotor experiments have also been reproduced by CFD for a range of wind angle, gap distances (d) and spin ratios. The EARSM model is used for the rotor-rotor interaction simulations at model scale. Figure 5 shows the results for a gap distance of 3*D with the spin ratios from both rotors set to 2. This condition shows the largest rotor-rotor interaction in the experiments.

The results are presented for two rotors, A and B. The rotors are placed behind each other with A being the forward rotor. The results are presented as the difference with respect to a standalone rotor. It is concluded that the interaction effects are generally well captured by the numerical approach. The changes in lift of the upwind rotor as well as the drag from rotor B in downwind configuration are particularly well reproduced. The trough at a wind angle of 15 degrees is observed to coincide with the aft rotor being in the wake of the front rotor as shown in Figure 6. This effect is further discussed in the full scale data presented in Paragraph 5.2.

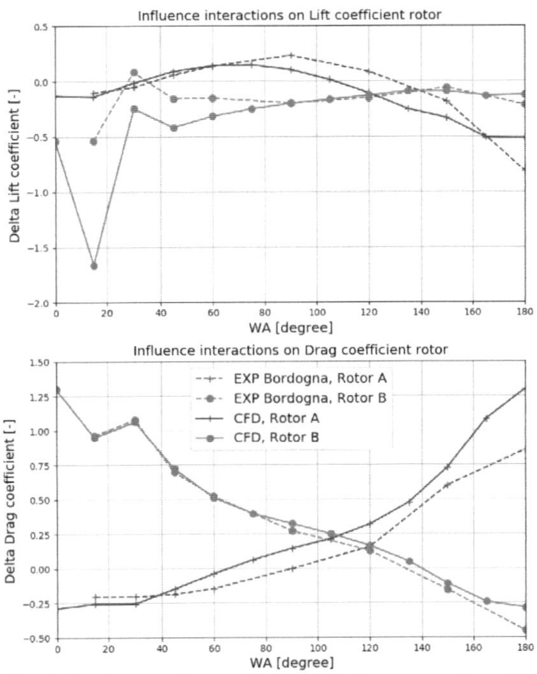

Fig. 5: Twin rotor interaction effects from experimental data [4] and CFD. Data at Re=1.05E+5. Both rotor at SR=2.0. Delta computed w.r.t to standalone rotor.

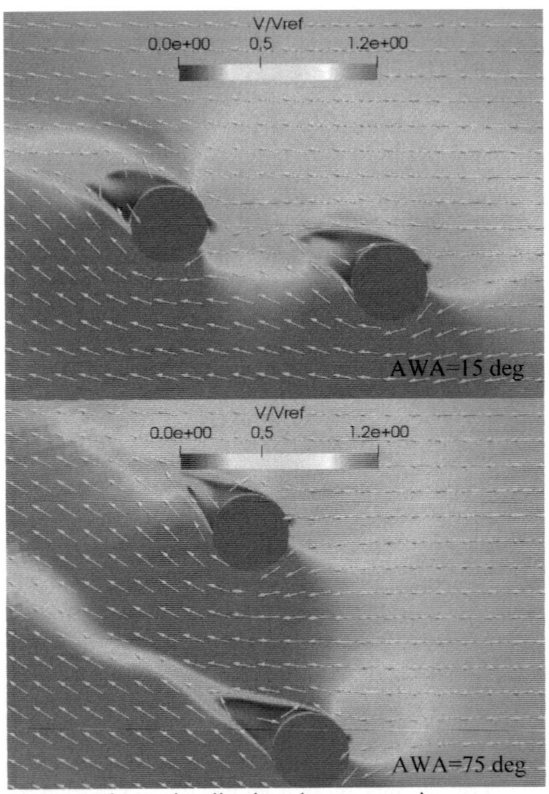

Fig. 6: Flow visualization between twin rotor at d/D=3.0. Data at Re=1.05E+5. Both rotor at SR=2.0.

4.4 Sensitivity of rotor-rotor interaction to the rotor support structure

The rotor-rotor interaction described in the previous section is also investigated at full scale with and without a realistic support structure underneath the rotor. The simulations are conducted with the kω-SST turbulence model since recent findings indicated that it yields better results than the EARSM model at full scale. The gap distance is kept equal to d/D=3.0 with SR for both rotors at 2.0. The data without support structure will be further described in section 5.2.

In this section the sensitivity to the support structure of the rotor is discussed. This is presented in Figure 7. It must be acknowledged that the full scale data are obtained with a symmetry plane at the bottom of the "unlimited" CFD domain while the model scale data are obtained within the numerical test section corresponding to the wind tunnel of Politechnico Di Milano. The comparison is judged acceptable since the interactions with the wind tunnel wall is assumed to be negligible at d/D=3.0 and the rotor is standing 10cm above the ~20cm thick boundary layer from the tunnel floor.

It can be observed the rotor-rotor interaction increases with the Reynolds number. The upwash and downwash effects from the rotors seems to be stronger at full scale Reynolds number. Moreover it can also be observed that the support structure from the rotor, modelled as a square pedestal of 3 m height, strongly influences the performance of the standalone rotor as well as the interaction. This is in particular visible for 30≤AWA≤150. For AWA≥150 and AWA≤30 the influence of the support structure appears to be negligible. Around AWA=90 deg a sudden change in trend is found with the support structure. Looking at the flow field it appears that this is caused by a large reversed flow region occurring around the support structure of the forward rotor. It is concluded that the support structure from the rotor should be included in when determining the interaction effects as its influence is non-negligible.

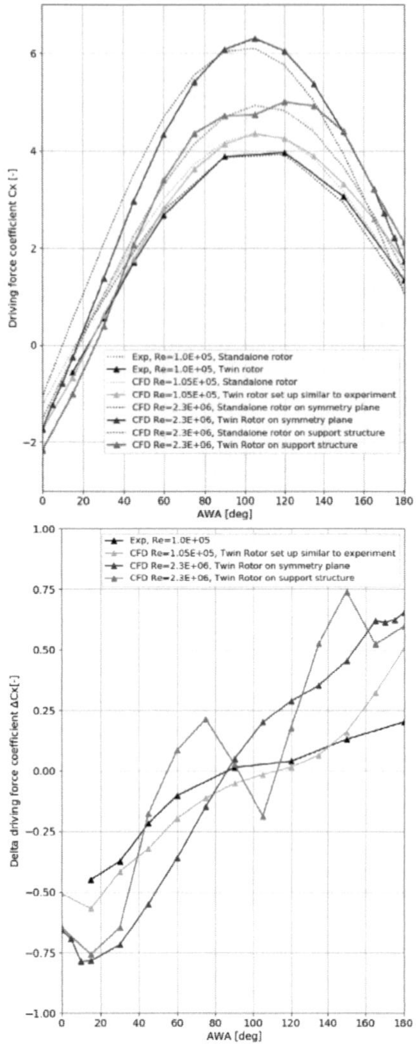

Fig. 7: Evolution of the driving force C_X due to rotor-rotor interaction computed by the CFD at model and full scale. The influence of the support structure on the rotor is shown at full scale. The change of driving force is also shown relative to the case without interaction.

4.5 Sensitivity of the superstructure interaction

The effect of the superstructure on the rotor is studied for a single rotor at the middle position on the MHTC under beam wind condition. The results are presented in Figure 8. The used uniform wind speed is equal to the wind speed at the middle of the rotor in the ABL. As the thrust coefficient differs significantly it can be concluded that the ABL should be included in calculations to determine the interaction with the superstructure.

As visible in Figure 9 the vessel creates a large flow obstruction which extends over the deck and part of the Flettner rotor and induces a loss of lift generation on the lower half of the rotor. These findings are in line with the results reported in [7].

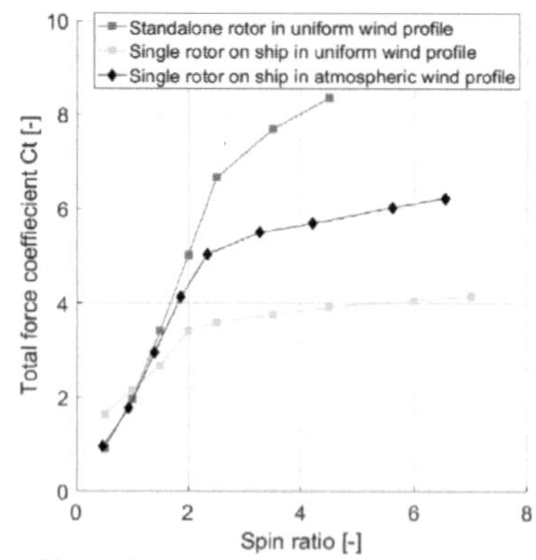

Fig. 8: Effect of the atmospheric boundary layer on the rotor performance.

It is also concluded that using an uniform wind profile is not recommended for this type of study as it artificially increases the blockage of the vessel compared to the ABL profile. It can be seen that the region of low speed is much higher in the case of uniform flow. In the situation with superstructure the effective wind speed seems reduced from that of the standalone rotor, leading to a significantly reduced thrust coefficient. The thrust coefficient is determined using the undisturbed wind speed at the middle of the rotor height which is equal for the three shown cases.

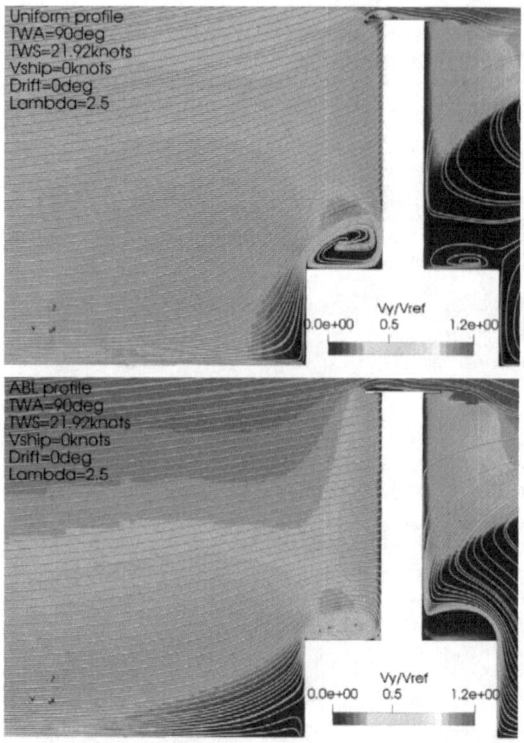

Fig. 9: Dimensionless flow around the superstructure. The flow is from left to right. On the top uniform inflow is shown and on the bottom inflow with an atmospheric boundary layer.

5. Application of the interaction model

Using the described numerical setup additional simulations with one Flettner rotor on the deck of the MHTC have been performed to confirm the independence of the spin ratio and twist on the disturbed wind effect from the superstructure and to describe its effect over wind angles. Secondly, the results of rotor-rotor interaction are described for a generic rotor configuration in full scale. Finally, the combined interaction of the superstructure and three rotors on the MHTC in operational condition is shown using calculations of this configuration for various wind angles.

5.1 Effect of the superstructure interaction

The effects of the superstructure on the performance of the middle rotor are studied including the ABL inflow profile for various apparent wind angles in the situation without ship speed. In addition two calculations are performed in an operational condition including forward speed resulting in twist in the wind profile. Using the calculated forces of the Flettner rotor the effective wind speed and angle is calculated. Using Equation (2) and Equation (3) the interaction coefficients for the disturbed wind are obtained and shown in Figure 10.

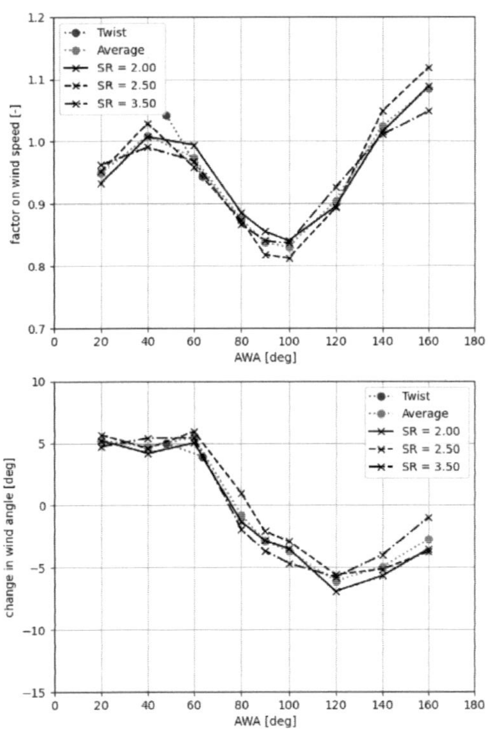

Fig. 10: Interaction coefficients for the disturbed wind from the superstructure for the rotor at the middle position based on the CFD calculations.

It can be observed that the effect on the wind speed and angle is significant, shifting the wind more to the beam in both bow and stern quartering apparent wind angles. At beam wind conditions the effective wind speed is reduced by around 20%. It can also be observed that the interaction is rather independent of the spin ratio which confirms that this simplification in the modelling framework is valid. Finally, it is observed that the effect of twist on the driving force is not significant, at least not for Flettner rotors. This makes it possible to use the defined apparent wind speed and angle at 10m height to determine the superstructure interaction instead of using combinations of ship speed and true wind speed as otherwise would be required. This reduces the number of calculations needed to calibrate the model.

The calibrated superstructure interaction model, using the average line over spin ratio's, is used to show the effect of the superstructure on the driving force of the rotors. The influence in terms of dimensionless driving force and change of driving force is shown in Figure 11, again made dimensionless using the undisturbed apparent wind speed at the height of 10m. The effect of the height of the rotor in the ABL is also included in these curves.

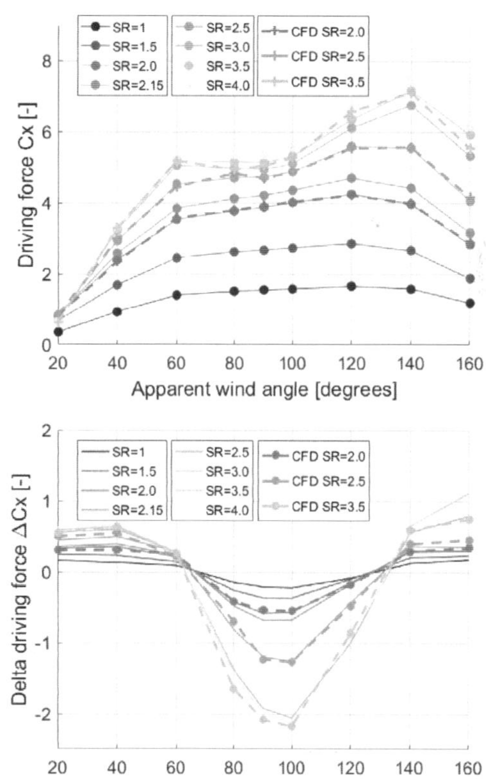

Fig. 11: Evolution of the driving force C_X computed by the superstructure interaction model and CFD for a range of spin ratios and apparent wind angles. The change of driving force is also shown relative to the case without interaction.

The results are presented for a range of apparent wind angles and spin ratios of the rotor and compared to CFD results. The spin ratio is here also defined using the undisturbed apparent wind at a height of 10m. It is concluded that the model correlates well with the CFD results. As discussed in the previous paragraph, it can be observed that a large reduction of driving force, up to 30% reduction for SR≥3, can be expected in beam wind conditions. This reduction was also visible in Figure 8 while comparing the red and black curves. On the contrary, in bow and stern quartering wind an increase of about 10-15% in driving force is predicted. It was found that this is mainly explained by the wind shift more to the beam in bow and stern quartering conditions even though the effective wind speed is also slightly increased. Arguably the ship will be sailing most of the time with apparent bow quartering (close hauled) wind due to the ship speed, yielding a positive influence of the superstructure interaction.

5.2 Effect of the rotor-rotor interaction

The effects of the rotor-rotor interaction are evaluated separate from the superstructure interaction using full scale CFD simulations of a generic twin rotor configuration. The simulations are conducted with the kω-SST turbulence model since recent findings indicated that it yields better results than the EARSM model at full scale. The rotor geometry from [4] is used meaning that the aspect ratio differs from the MHTC, namely AR=5 instead of 6, and that the end disk ratio is slightly larger. The vessel is not included in the simulations. Instead a symmetry boundary condition is applied at the foot of the rotor, representing ideal flow although it might not be realistic for some configurations. The two rotors are positioned at a gap distance d, defined centre to centre, corresponding to d/D=3 and d/D=7.5. Two spin ratios are investigated: SR=2 and 3.5 for d/D=3 and SR=2 for d/D=7.5. Both rotors rotate at identical rotational speed.

It can be observed in the using the calculated forces obtained interaction coefficients shown in Figure 12 that the forward rotor (A) experiences an increased inflow angle due to the upwash effect. For the aft rotor (B) the inflow is significantly more from the bow due to the downwash effect. Also the average is shown, giving the effect on the configuration of the two rotors combined. This explains the effect of the interaction on the driving force of the two rotor configuration. The overall (average) effect is that the inflow is a rather constant 5 degrees closer to the bow. In addition a decrease in the effective wind speed at the aft rotor is observed at the wind angle of 15 degrees where the aft rotor is in the wake from the forward rotor. This is illustrated in Figure 13.

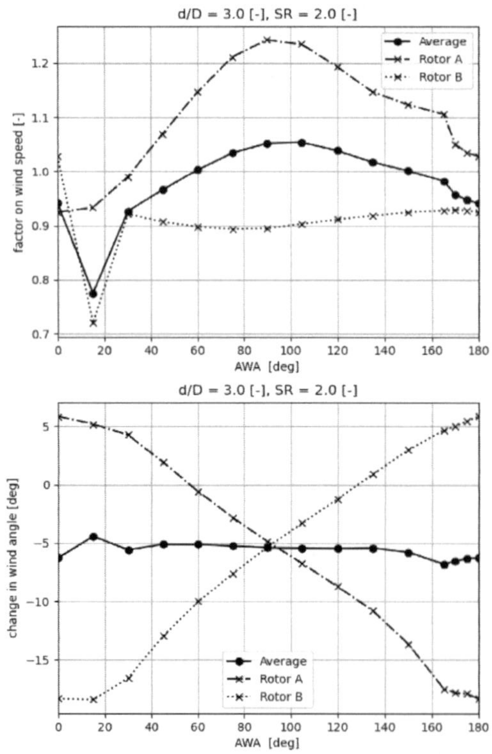

Fig. 12: Interaction coefficients determined for the two rotor configuration

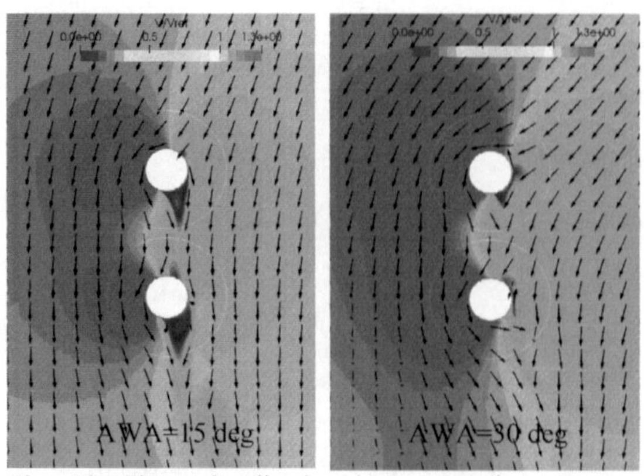

Fig. 13: Flow visualization between twin rotor at d/D=3.0. Data at Re=2.3E+6. Left: rotors at SR=2. Right: rotors at SR=3.5

The effect of the spin ratio on the interaction coefficients is shown in Figure 14. It is observed that indeed the coefficients depend on the spin ratio as was expected. It seems that the aft rotor is at an increased wind angle in the wake of the front rotor for the higher spin ratio. This can be explained by an increased downwash effect associated the larger circulation of the forward rotor as visible in Figure 13.

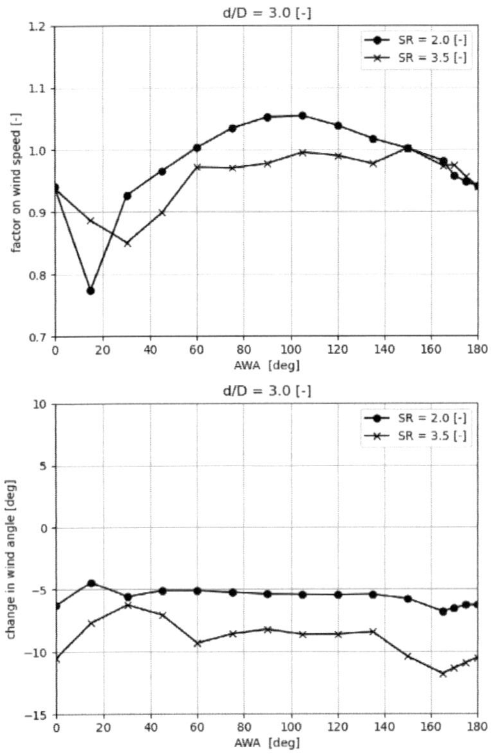

Fig. 14: Effect of spin ratio on the average interaction coefficients for the configuration of two rotors

The effect of the interaction on the driving force is shown in Figure 15. Looking at the combination of both rotors, a negative effect on the driving force is predicted in bow quartering wind while the driving force of the configuration increases in stern quartering winds. As anticipated the effect of rotor-rotor interaction on the driving force increases with increasing spin ratio and reduces with increasing gap distance. The increase in driving force of the front rotor (A) is roughly antisymmetric to the reduction in driving force on the aft rotor (B).

In bow quartering wind the forward rotor benefits from the upwash from the aft rotor. The aft rotor on the other hand shows a decrease in driving force due to the increase apparent wind angle from the downwash of the forward rotor. This effect increases at higher spin ratio. The flow phenomena is identical in stern quartering wind but the benefits in terms of driving force are reversed. This effect can be seen in the interaction coefficients shown in Figure 15 and is illustrated per rotor for SR=2 d/D=3 in Figure 16

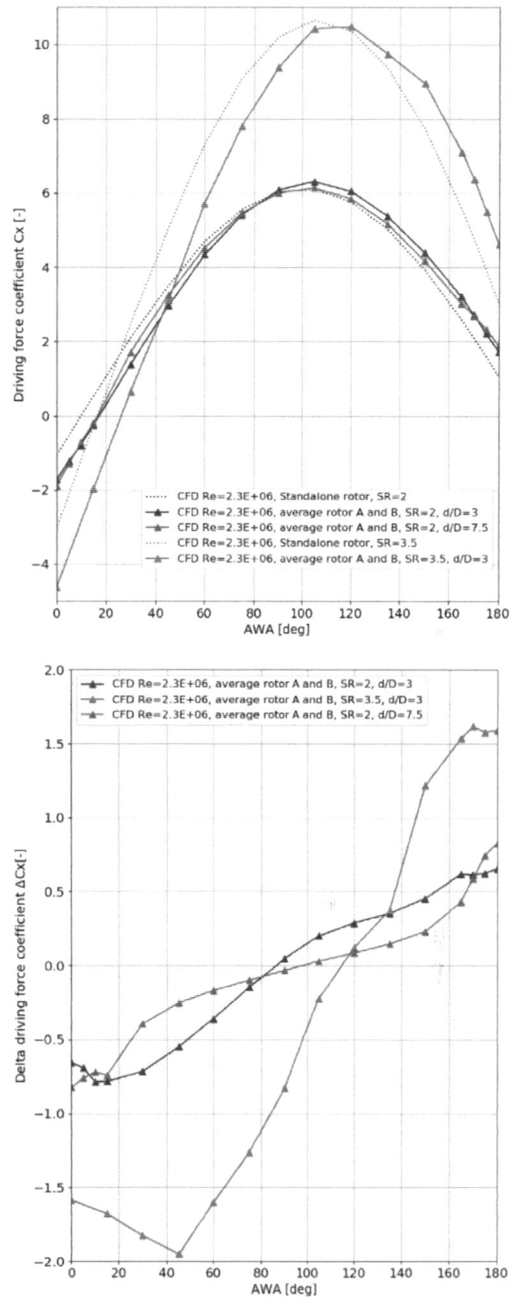

Fig. 15: Evolution of the driving force C_X due to rotor-rotor interaction computed by CFD. The change of driving force is also shown relative to the case without interaction.

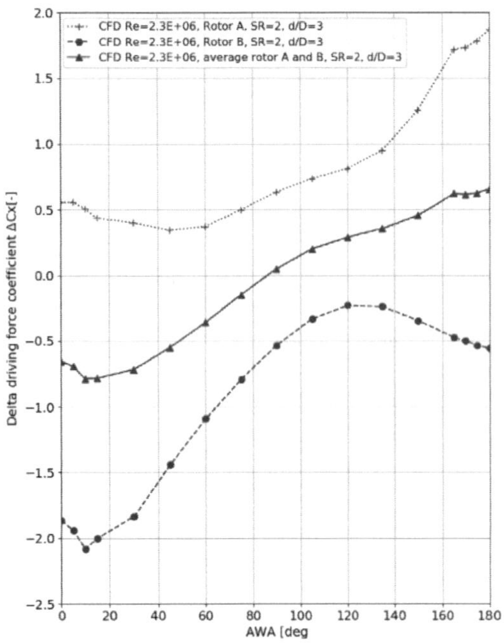

Fig. 16: Evolution of the change in driving force ΔCx due to rotor-rotor interaction computed by the CFD for SR=2 d/D=3. The change of driving force is relative to the case without interaction.

5.3 Effect of the combined interaction

The effects of the combined interaction are evaluated based on dedicated CFD simulations with three rotors on the MHTC, thus including all interaction. These results are compared to the effect of the superstructure based on the derived interaction coefficients for the location of the middle rotor, thus excluding rotor-rotor interaction. This is presented in Figure 17. It should be noted that some difference in superstructure interaction could be present due to the different location of the fore and aft rotor. Each point corresponds to a specific sailing condition determined by a PPP simulation: combination of spin ratio and true wind angle at a given wind speed and ship speed as shown in Table 2. Effectively the spin ratio ramps up from SR≈1 to SR≈4.5 with increasing apparent wind angle. Each rotor runs at identical rotational speed. The rotor still keeps running in head wind to decrease drag, although depending on the power consumption of the rotor it might be stopped as well.

Table 2: Operational conditions used for the configuration with three rotors on deck.

TWA	TWS	V_S	$N_{Flettner}$	β
[deg]	[m/s]	[kn]	[rpm]	[deg]
0	10.5	12.0	94.8	0.0
20	10.5	12.0	114.6	2.0
40	10.5	12.0	145.7	3.1
60	10.5	12.0	174.4	4.0
80	10.5	12.0	198.5	4.2
100	10.5	12.0	210.5	3.5
120	10.5	12.0	205.5	2.1
140	10.5	12.0	178.8	0.6
160	10.5	12.0	143.6	-0.4
180	10.5	12.0	43.4	-0.3

It is concluded from Figure 17 that the results from the configuration with three rotors on deck and the modelled superstructure interaction indicate similar driving forces in a large portion of the wind conditions, 30≤AWA≤100. Since the superstructure interaction model does not include the rotor-rotor interaction this suggests that the disturbed flow due to the superstructure is dominant over the rotor-rotor interaction in this region and the effect of the location of the fore and aft rotor might be neglected.

The deviation in the curves indicate that rotor-rotor interaction does affect the results at other wind angles. Following from the previous section it is expected that at AWA≈90deg the rotor-rotor interaction has a limited to neutral influence on the total driving force of the rotors, see Figure 15. The rotors have a rather large distance between them of d/D≈9.5. For smaller apparent wind angle, 30≤AWA≤60 the spin ratio varies between 1.3≤SR≤2. Considering that the gap distance on the MHTC is large and that the spin ratio is lower than 2 it can be assumed that rotor-rotor interaction is still limited for 30≤AWA≤60.

At AWA≤30 the CFD results show a smaller driving force than results from the superstructure interaction. This is supported by the findings from Figure 15. The rotor-rotor interaction induces a reduction of the driving force in bow quartering wind. However at stern quartering wind, AWA>120 with SR>4 the driving force with only the disturbance of the superstructure is higher than obtained from CFD for the configuration with three rotors on deck. This while rotor-rotor interaction should be beneficial in this region. It is unclear what is the reason behind this mismatch. It might be caused by the dependence of the superstructure interaction model to the position of the rotor on the ship for stern (and bow quartering) wind angles. The superstructure interaction is determined for the middle rotor and may differ at the fore and aft rotor location.

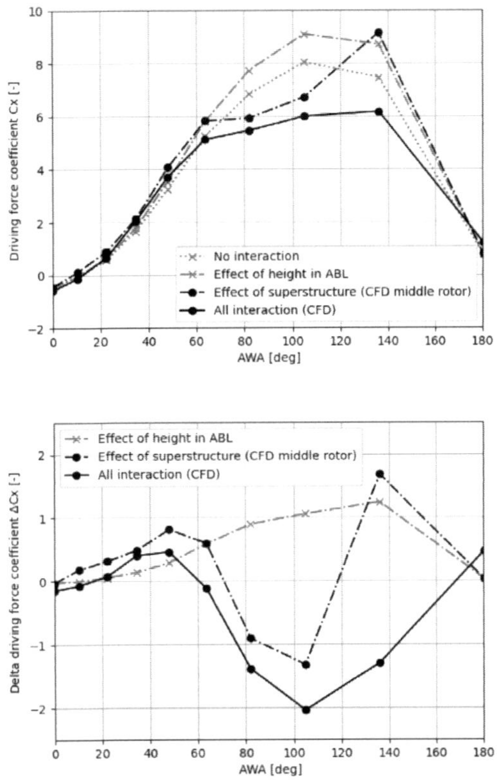

Fig. 17: Evolution of the driving force Cx computed by the interaction model and CFD for a range of sailing condition. Each sailing condition corresponds to a specific combination of AWA and SR. The change of driving force is also shown relative to the case without interaction.

6. Conclusions and future work

The wind propulsor interaction effects have been investigated for the MARIN Hybrid Transition Coaster (MHTC) in this paper using RANS calculations. A simplified modelling framework decomposing interaction effects into the summation of multiple interactions has been proposed. The assumptions required to make the simplifications in the model were validated where possible. Such a framework is required for including the interaction effects in PPP predictions.

The RANS numerical set up has been extensively verified and validated using different experimental data including a twin rotor configuration. A good correlation is found with experiments for both standalone rotor and twin-rotor configuration. Validation of the effect of the disturbed flow due to the superstructure has not yet been performed due to lack of experimental data. A sensitivity study showed that including the atmospheric boundary layer is necessary as a constant wind profile would results in excessive blockage from the vessel which exaggerates the interaction. A representative description of a foundation pedestal also appears to be important, both for the standalone rotor performance as the rotor-rotor interaction.

The study is performed on a practical test case: the MHTC concept with three Flettner rotors. It was found that the proposed model could capture the effect of the disturbed flow from the superstructure rather well and that the effect of the operating setting of the wind propulsor and twist due to forward ship speed could be neglected. Simply taking a point sample for wind condition and developed forces at the mid height seems sufficient for Flettner rotors, but may require further work for other wind propulsors with a stronger dependence on angle of attack. It was observed that the superstructure interaction is dominant for near beam wind. For bow quartering wind the effect of rotor-rotor interaction is also important. The results in stern quartering wind indicate that also the interaction of the superstructure might need to be determined for the various positions of the rotors, which should still be confirmed by future research.

For a generic case not representing the MHTC configuration, rotor-rotor interactions have been investigated by RANS simulations. Large interactions have been found. It is concluded that in a twin rotor configuration the interaction increases with increasing spin ratio and decreasing distance between the rotor as expected. The interaction yields a reduction of driving force in bow quartering wind and an increase in stern quartering wind. In beam wind the interaction tends to cancel out. It is confirmed using these variations that the interaction coefficients depend on the operating condition of the wind propulsors as well, contrary to the interaction with the superstructure.

Finally it is recommended to continue the study on multiple aspects. The independence of twist (and indeed generally in PPP calculations) should be reconsidered for other types of wind propulsors. Further, it is observed that the rotor-rotor interaction can be significant and cannot be omitted also at rather large distances between them. The challenge here is that a large matrix of calculated conditions is required to include these in a PPP. It might therefore be required to use a lower fidelity method such as numerical lifting line methods. Ongoing research should show if the trends obtained using such a method are accurate enough for this purpose. Lastly, validation material including superstructure interaction is required to further validate the numerical RANS approach.

7. ACKNOWLEDGEMENTS

This research is partly funded by the Dutch Ministry of Economic Affairs.

8. REFERENCES

[1] ABS, 'Guide for Building and Classing Offshore Wind Turbine Installations', *ABS*, 2014.

[2] DNV, 'Environmental conditions and environmental loads', *DNV*, 2010.

[3] GARENAUX, M., SCHOT, J. J. A. AND EGGERS, R., 'Numerical analysis of Flettner rotors performances on the MARIN Hybrid Transition Coaster', *HIPER*, 2020.

[4] BORDOGNA, G., 'Aerodynamics of wind-assisted ships: Interaction effects on the aerodynamic performance of multiple wind-propulsion systems', *doctoral dissertation,* 2020.

[5] SPALART, P. R. AND RUMSEY, C. L., 'Effective inflow conditions for turbulence models in aerodynamic calculations', *AIAA, vol. 45, no. 10, pp. 2544-2553*, 2007.

[6] DA-QING, LI., ANDERSEN, M. L. AND ALLENSTROM, B., 'Performance and vortex formation of Flettner rotors at high Reynolds numbers, *29th Symposium on Naval Hydrodynamics* , 2012.

[7] JONES, L., PRINCE, M., HUDSON, D., COCKS, J., 'Predicted fuel-savings for a Flettner rotor assisted tanker using computational fluid dynamics', *RINA International Conference on Wind Propulsion*, 2019.

9. AUTHORS BIOGRAPHY

M. Garenaux holds the current position of project manager at MARIN in the specialized vessel team. He is responsible for numerical studies on various type of vessels including special vessels. He focusses in particular on propeller-hull interactions and energy saving devices. Together with other colleagues at MARIN he is also actively working within the wind propulsion research team.

J. J. A. Schot holds the current position of project manager at MARIN in the transport and shipping team. He is responsible for numerical studies supporting the design of cargo vessels with a focus on emission reduction, including wind propulsion. His previous experience includes work on the application of CFD in both a research and commercial setting.

RONDOUT RIVERPORT 2040: A COMPREHENSIVE PLAN FOR A WORKING WATERFRONT AND THE TRANSPORTATION OF GOODS AND PEOPLE IN A CARBON CONSTRAINED FUTURE

A Willner, The Center for Post Carbon Logistics, USA

SUMMARY

Rondout Riverport 2040 proposes a pragmatic, positive, and prosperous vision for the near future in which the communities of Kingston and Esopus, New York, are enriched by a transformed port, boasting a shoreline brimming with heritage and leading-edge maritime commerce and working waterfront technologies that profit and engage individuals, businesses and communities, allowing for an equitable transition beyond fossil fuels, where all work as together to forge a vital and vibrant economic bond with the greater Hudson Valley Bioregion.

1. INTRODUCTION

Rondout Riverport 2040 offers the communities of Kingston and Esopus, New York, a visionary template and extraordinary opportunity for remaking and transforming the Rondout Creek and Hudson River Working Waterfront over the next 20 years.

Rondout Riverport 2040 provides a trailblazing and sustainable development guideline for our community, harnessing and enhancing our region's widely shared prosperity, even as we enter an economically demanding carbon-constrained future.

The Riverport, in 2040, as envisioned here, will offer far more capacity, while being significantly more compact in land area, more robust, and resilient than the current patchwork of diffuse land uses found on today's waterfront. The core mission of tomorrow's port is the post carbon maritime transport of goods and people up and down the Hudson River and beyond.

The Riverport is designed to attract shipping, distribution, commerce, hospitality, and craft businesses, creating a dynamic collaboration and nexus for optimized local and regional market productivity. The result: an economically, culturally, and environmentally resilient post carbon working waterfront – a gateway to the Hudson Valley and the World.

Rondout Riverport 2040 is a signature project benefiting from the creative contributions of its founding partners, The Hudson River Maritime Museum [1], The Schooner Apollonia [2], Sustainable Hudson Valley [3], The Center for Post Carbon Logistics [4], and many additional partner organizations, local governments, and institutions to address and transcend the near future threats of sea level rise; increasingly turbulent and extreme weather events; and unexpected global, national and regional economic shocks. Rondout Riverport's versatility will depend on the linking of its economic opportunities with environmental restoration and sustainable commerce. Embracing this multi-generational project will also be a source of inspiration for broader long-term action on climate change.

We can best accomplish these visionary waterfront goals via an integrated "placemaking" approach.

Placemaking provides a method for answering critical questions: What are the best ways to mobilize and coordinate our many community assets? How do we effectively draw on public and private partnerships to creatively identify opportunities? How can we successfully coordinate our implementation efforts? And where do we find the resources needed to accomplish our vision for a transformed riverport?

We do not have to wait until 2040 to start benefiting. Communities can begin now, as they participate in a vigorous planning process, while taking key steps for future proofing our shoreline against the harms threatened by a more politically, economically, and environmentally chaotic planet in a post-carbon future. The path to a bright, sustainable future starts with community engagement and data collection to build an actionable vision for the Rondout Riverport, a vision that incorporates a proud sense of community and place, local stewardship, and widely shared economic opportunity. The choice is ours.

2. A VISION FOR RONDOUT RIVERPORT WORKING WATERFRONT, CIRCA 2040

Imagine: It is a hot, late autumn day along the Hudson. From the rooftop of a trading house[1] in Kingston, a ship spotter sees the topmast of a large sloop. The sloop signals a waiting solar electric steam tug, the *Augustin Mouchot* [5], which tows the engineless sailing ship toward a berth in the newly completed Rondout Inner Harbor.

The sloop, the *Pete Seeger*[6], is loaded with high-value cargo from abroad, transferred in New York Harbor from the oceangoing post carbon Eco-Clipper [7] *Jorne Langelaan*.[8] The mixed freight consists of Caribbean fair-trade coffee and cocoa beans bound for the Hudson Valley's

1 A trading house is an exporter, importer and a trader that purchases and sells products for other businesses. Trading houses provide a service for businesses that want international trade experts to receive or deliver goods or services.

roasters and chocolatiers, along with preserved tropical fruits and rum destined for local Kingston storehouses.

The *Seeger's* crew put a harbor furl on the hemp cloth sails, even as other crew members ready the on-board cargo gear. The sailors open hatches and set up the cargo boom which will do most of the heavy lifting. The crew can also access the Rondout's floating cargo cranes for heavier or bulkier freight.

Figure 1. Unloading Black Seal in Brooklyn

These locally trained young seafarers are in good spirits, looking forward to spending some time ashore, and to a few drinks of locally made brew, cider, and spirits. Like any sailors, they are also hungry and ready for a good meal at a waterfront tavern – the local fare includes dishes harvested from the Hudson's new artisan fishery and from oyster beds seeded in shallows created by collapsed former piers and abandoned roads, submerged downriver over the last 20 years due to rising seas and river levels. After dinner, the sailors walk along the sea-life-encrusted seawall, built from repurposed concrete and stone from former waterfront roads, buildings, and piers inundated by the Hudson's rising waters.

A longshore crew, warehouse workers, drovers, and their electric-assist people-powered tricycles and wagons, converge at the waterfront's new storm-proofed floating dock – which rises and falls with surging tides. Cargo surveyors assist with the unloading of the coastal Schooner, *Sam Merrett* [9] down from Nova Scotia with a load of lobsters. The square foremast tops'le Schooner *Kevin Kerr Jones* [10] is unloading citrus from Savannah. Other stevedores are loading the solar electric Feeny [11] shipyard-built canal barge *David Borton* [12], bound for ports up the Hudson River and through the Erie Canal with a destination at Buffalo. Some smaller solar barges are loading for Port Jervis, on to the newly opened Delaware and Hudson Canal [13].

The (s)low tech Rondout Riverport is modern and efficient. The port is no longer dependent on prohibitively expensive fossil fuels, nor the notoriously unreliable overseas energy supply chain. Instead, Rondout makes the best use of old and new – tried and true 19th century technology blended seamlessly with 21st century solar and battery electric gear and vehicles. More people are at work today on the waterfront than at any time since the 1920s; there are more warehouses and trading houses, shipbuilding, repair facilities, and docking facilities than at any time in the Rondout's nautical history.

Just behind the waterfront are coopers [14], using sustainably harvested local oak; sail and ropemakers, utilizing New York hemp; forges and foundries using concentrated solar [15] to form bronze fittings. Riggers are hard at work in ropewalks [16] making running rigging and dock lines to equip the numerous commercial and recreational sailing ships and boats.

Figure 2. Rope Walk

Dry docks and shipyards look out on bikeways and walkways circumscribing the tidal flats, from which hundreds of locals and tourists observe the port activity, safe in the knowledge that food and goods continue to pour into a port that – thanks to good planning 20 years ago – is admirably adapted to keep pace with a changing climate and evolving post carbon economy. All of this could be, if only we take a can-do proactive approach toward tomorrow.

Figure 3. Solar Electric Canal Barge

3. REINVIGORATED WATERWAYS: THE FOUNDATION OF A RESILIENCE STRATEGY

"Contrary to the techno-paradise that some expect, my belief is that our future will likely resemble our past, and that we may fall back on proven, low energy approaches to supporting human life that have been historically proven to work. "Isn't that pessimistic?" asked the interviewer. I replied that I don't think so. It is in my view even more pessimistic to imagine a world continuing on the current path, becoming a place in which there is no place for human labor or creativity, where rather than doing things with our backs and hands and minds, we must instead wait passively for conveniences and solutions to be marketed to us. That, to me, is the most depressing future imaginable." – Erik Andrus [17] Founder the Vermont Sail Freight Project [18]

Not so long ago, in the 19th and early 20th centuries, the Hudson River bustled with commerce and lay at the heart of a thriving network of marine highways linking the largest cities and smallest communities to a web of regularly scheduled transportation routes – waterways stretching from the Atlantic west to the Great Lakes.

Boats of all sizes served local cargo and passenger needs. Schooners, sloops, barges, and steamboats connected river town inhabitants. Farmers, merchants, and oystermen relied on this vibrant and diverse fleet to deliver goods to market and to bring back supplies.

The Hudson River – and the ships and boats sailing her – were vital and integral to those who worked, lived, and thrived along our inland waters, putting places like Kingston and Esopus on the map.

Historically, thousands of vessels plied these marine highways, sailing up and down the Hudson Valley, delivering fresh local farm produce ranging from apples to applejack, fish, and shellfish, and carrying passengers to ports along the way.

Figure 4. View near Anthony's Nose [19]

The Kingston and Esopus' Rondout Creek and Hudson River working waterfront has long been a key contributor to our region's financial wellbeing – though it could be so much more. Now, as we enter a carbon constrained future our Riverport is poised for rebirth, to again become a key regional hub for the transport of goods and people.

As we move into a world facing increasingly tough political, economic, and environmental challenges, we must ask ourselves: How shall we, living beside the Hudson River, meet the looming threats of climate change, rising sea level, aging infrastructure, changes to global shipping patterns, threats to food security, upheavals in energy production and distribution, and the risks all these disruptions could bring? As in the 19th and 20th centuries, the answer to those questions, and the solutions to our problems lie only as far away as the lapping waters of our home river.

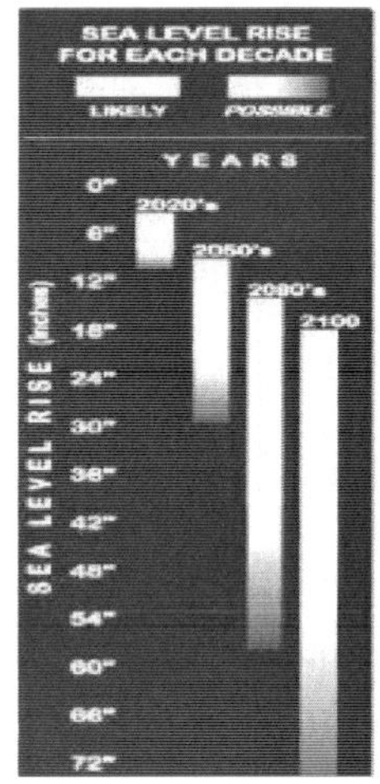

Figure 5. Scenic Hudson Sea Level Rise by Decade

Rondout Riverport 2040; A Comprehensive Waterfront Plan for a Working Waterfront and Transportation of Goods and People, offers a pragmatic look forward to what – with proper preparation, cooperation, and investment – could result in a revitalized and highly profitable Rondout Riverport at mid-century. This plan provides a practical salient vision of resilient shorelines [20] and a working waterfront [21], redesigned to protect our community from sea level rise and storm surges [22], built to accommodate a wide spectrum of business, cultural and social uses that will benefit our communities and the Hudson Valley Bioregion. This, to put it simply, is a waterfront proposal that "floats all boats," promising equity and prosperity for our citizens, large and small industries, impact investors [23], entrepreneurs, craftspeople, environmentalists, boatmen and women, dreamers, and doers.

But here is a warning: An optimistic future depends on our will to make it so. If we pursue politics and policy as usual, we could face a grim tomorrow as our region is hobbled by climate change: Abandoned, flooded, mouldering shoreside buildings and piers; low-lying and failing sewage treatment plants and electric utilities; eco-refugees crowding our upstate communities seeking limited food and shelter; and a polluted, dead estuary as oil and chemical plants are inundated. Despite sincere efforts at incremental change and adaptation planning, without visionary action right now, our region could face a dire tomorrow marked by rising water and plummeting economic fortunes. The choice remains ours.

The reality of escalating climate change makes clear that we must redesign our economies if we are to maintain quality of life in a carbon constrained future. A major opportunity offers itself: to take advantage of our wealth in waterways and return to our bioregion's nautical roots and pioneer a new industry grounded in tried-and-true technology that once drove our economy: low/no carbon shipping, and post carbon transportation businesses, and organizations [24].

In the New York City metropolitan area today, 80% of freight transport is carried by truck [25], a mode of transportation that is congesting our highways, increasing air pollution, and entirely dependent on fossil fuels. In a carbon constrained future, sustainable water transport (an innovative mix of sailing vessels, hybrid/fossil fuel-free electric ships, and people/electric powered transport) will almost certainly be a necessity.

As the climate crisis deepens, water-based transportation routes can link communities and promote resilience throughout the Hudson Valley – doing so without congestion, without pollution, while being energy efficient, non-dependent on increasingly expensive fossil fuels, and potentially very profitable.

Water-based transportation, once ubiquitous on the Hudson, is just about the only form of transportation, other than the bicycle, that requires little or no roadway maintenance. Navigation channels are less costly than roads to keep up; they do not require a large industrial base and are far less energy-intensive than alternatives.

We need not look far for proof: The 363-mile-long Erie Canal system [26] linking the Atlantic Ocean with the Great Lakes, has been continuously operational since 1825. The cost of keeping it running is tiny compared to that of equivalent highway mileage. The Hudson and its linked waterways comprise the greatest set of transportation assets in the world – assets greatly underutilized today. Those "blue highways" will see their status grow in a post carbon world, and communities along them will prosper as a result.

Kingston and Esopus are two such communities. The Rondout Riverport is strategically located to be part of this great renaissance: located just ninety miles from one of the greatest ice-free harbors on earth; and sixty miles from the entrance to the Erie Canal.

But to make this opportunity a reality, Riverport infrastructure must be created to increase capacity, while being nimble enough to respond to rising sea and river levels and worsening storm surges, as well as shifting economic tides. The port will also need to be made accessible to smaller, more numerous vessels on a protected and restored working waterfront [27]. To thrive as a maritime and commercial center in a carbon constrained era, Rondout Riverport's infrastructure must include:

- Charging stations for electric and electric hybrid vessels, flood-proof storage and production facilities for biofuels [28] like methane (produced by sewage treatment plants), biodiesel (from restaurant used fryer fat) [29], and hydrogen [30] (created from seawater while sailing vessels are underway); A flood-proofed waterfront and flood-proofed [31] warehouses and trading houses.

- Local ship and boatbuilding and repair facilities to support our local commercial fleet [32].

- More traditional break bulk [33] cranes for transfer of palletized, and bagged cargo [34]

- Across-docking facilities for transfer of goods from ship-to-ship and from ship-to- first-and-last-mile providers (i.e. small sailing, rowing, hybrid vessels as well as people/electric powered commercial trikes [35], and wagons);

- Access to innovative training facilities [36] to provide a labor force: the new traders, river rovers, seafarers, and port workers. This labor force will need training based on models for "preserving the tools and skills of the past to serve the future."

Rondout Riverport will not stand alone, but will be integrated into the greater Hudson Valley Bioregion, along with the wider Northeast and U.S. transportation and distribution system, with which it will engage collectively and creatively to unleash an extraordinary historic transition to a future beyond fossil fuels [37]; a future that is vibrant, abundant, resilient, and ultimately preferable, more equitable, and more economically viable than the current model.

Rondout Riverport 2040 will serve as an empowering example to our bioregion and our country – demonstrating the viability of ethical livelihoods and teaching beneficial sustainable technologies that do minimal socio-environmental harm; methodologies that foster self-reliance and promote Slow Tech hands-on work practices.

Figure 6. Caulking Seams

The result: entrepreneurs, professionals, technicians, craftspeople, academics, and students from across our bioregion, and across the United States, will be drawn to our state-of-the-art waterfront – gathering here to learn from each other. Our waterfront will be like no other in our region, or maybe in the nation: Becoming a *living laboratory,* cultivating not only practical and sustainable energy, commerce, and transportation solutions, but generating a flow of fresh, pathfinding ideas.

To bring these advanced infrastructure changes about – working with partners throughout the region – we will need to establish: a new binding agreement with the region's farmers and farm advocacy organizations in our "foodshed" that offers subsidized support of infrastructure including, but not limited to:

○ Full employment in year-round growing season zero carbon greenhouses [38].

○ First and last mile transportation of agricultural products to processors and the waterfront (using existing and new rail-trails as bike/trike corridors);

○ Solar powered cold storage at critical locations.

○ Year-round indoor farm markets.

○ An inter-port agreement with small and mid-sized ports along the Hudson River, the Erie and Champlain Canals, and New York Harbor:

○ This agreement would include the sharing of information on resilience and "future proofing" of all waterfronts.

○ Establish a *Sustainable Working Waterfront Toolkit* – making available the historical and current uses and economics of New York's waterfronts as a resource. The toolkit must include legal, policy, and financing tools that river ports and waterfront communities can tap into so as to preserve and enhance local and regional port facilities.

• Small ship access to flood proofed regional produce and fish markets [39].

○ The Hunts Point Market and Fish Market must be made accessible to small ships, delivering farm goods from upstate and returning with seafood from the Market.

• A new agreement with transport and longshore unions that allows ships to load and unload with their own equipment. Local industry will need to work in close conjunction with the unions to hire and train more people for post carbon longshore work.

• A partnership with the region's Maritime Academies, the Hudson River Maritime Museum's Wooden Boat School [40], and a new maritime trades high school based on the New York Harbor School [41] to train mariners, and to teach the logistics careers required to serve the new post carbon working waterfront; with the ultimate goal being the creation of a Training Center [36] in the mid-Hudson Valley where professional practitioners and apprentices can participate in practical workshops to relearn maritime and other heritage skills and old/new technology to serve present and future needs;

• An endowment for the preservation and utilization of traditional maritime skills and tools, the establishment of a traditional knowledge database/Wiki; library; and pre/post carbon tool, technology, and machinery collection. This innovative interactive educational resource serves to preserve, restore, and promote the re-use of traditional skills, integrating those skills and methods with modern know-how and appropriate post carbon technologies.

• Create maritime mixed-use zones where public parks, walkways and bikeways are built in flood zones and are adjacent to and part of the working waterfront – acting as a source of recreation and as a vital part of flood control.

• Establish a strong working relationship with NOAA's National Sea Grant Program for working waterfronts [42].

• Advocate for a reduced, less intrusive regulatory role for the US Army Corps of Engineers. Instead, encourage Corps funding be channelled into partnerships with other agencies, local non-profit organizations, and an engaged public to develop, and redevelop climate change-resistant and resilient Hudson River ports, and to create living shorelines, restored wetlands, and estuarine habitats resulting from the removal of bulkheads and restoration of intertidal habitats [43].

Through this diversity and combined effect of uses, Rondout Riverport's working waterfront will also:

• Create jobs in sailing, logistics, shipbuilding, harbor maintenance, craft, food production and more.

• Revitalize the waterfront community via economic development combined with better public access and recreation.

• Improve regional food production and distribution, linking producers to markets in the Hudson Valley and beyond.

• Design and build a maritime commerce micro-hub for aggregation, warehousing, co- packing, and marketing.

Rondout Riverport will be the homeport for future-proof sailing, alternative fuel, and solar electric ships. It will provide training in maritime skills, shipbuilding, and longshore trades, while also educating crews in "earth care, people care, and fair share" principles [44]. These future-proof [2] ships and their locally trained crews will carry *people, goods, and knowledge* to and from towns along the Hudson and on the region's canals.

2 The term "future-proof" refers to the ability of something to continue to be of value into the distant future – that the item does not become obsolete.

As Rondout Riverport becomes the working waterfront of tomorrow, the constraints, and advantages of smaller and (s)lower tech modes of transport must be considered in every aspect of the port's design [45]. Historic and modern technologies must meld seamlessly to offer approaches that are more self-sufficient and sustainable. Just one example: ships of all sorts, meeting a variety of needs, will have to be built (and rebuilt) locally, from locally sourced or recycled materials, and be crewed by locally trained seafarers. These new vessels will likely be different than the ones we build today; smaller, more versatile, adaptable, energy-smart, and affordable.

As fossil fuels become more expensive or less available, replaced by alternative sources, and are restricted by climate change policy – port infrastructure will need to be part of a carbon neutral trading network for "short sea shipping [46]" that links us to the region and the world, the Hudson Valley, the New York/New Jersey Harbor, coastal waterways, and transfer points for goods from overseas.

Moreover, the Rondout Riverport will be well positioned to become a laboratory for carrying cargo under sail [47] as public agencies [48], and private companies [49] accelerate their investigations of the potential economic and environmental benefits of transferring more cargo from roadways to waterways.

4. IMAGINE THE FUTURE, REALIZE THE VISION

Life at the water's edge is rapidly changing. The impacts of new technology, patterns of urban development, and globalization are redefining global logistics, and while some waterfront cities will thrive as ports and grow under these new conditions, others will need to evolve to survive and succeed....

The Rondout Creek today, lapping at the shores of Kingston, and of the Sleightsburg and Connelly hamlets in the town of Esopus, is in the flux of significant change. The waterfront as it is, represents an amalgam of positives and negatives. At its best, it boasts commercial shipyards, marinas, marine services businesses; institutions including the Hudson River Maritime Museum (and its Wooden Boat School, and Shipyard [32]); along with wetlands, open space, promenades, magnificent scenery [50] and recreational possibilities. But at its worst, it is marred by brownfields [51], combined sewer overflows [52], and a variety of non-water dependent [53] uses that make poor use of water accessibility, marine transportation, and port possibilities.

Most unfortunate of all: existing development plans lack a sweeping vision and often fail to take a future into account dominated by climate impacts, including severe storm surges, along with a steady sea and river level rise that will soon inundate portions of the currently existing Rondout and Hudson shoreline [54]. Plans that fail to take climate change into account will drown in insolvency.

Over the past few years, a variety of plans and proposals have been put forward, each with particularly good elements, but also with gaps and flaws:

1. The City of Kingston and town Esopus are working with stakeholders and partners on The Rondout Waterfront to improve the resiliency and sustainability of the shoreline, implement an economic development strategy, and cultivate better access to the river via waterfront parks and open space for people on foot, on bicycle, and launching boats.
2. Kingston's Weaving the Waterfront [55] planning process has many elements including: the Waterfront Resiliency Project, Rondout Riverport Shoreline Stabilization and Public Access Project [56], Kingston Point Park Improvements Project [57], Kingston Point Climate Adaptive Design [58], Kingston Point Rail Trail Phase 2 [59] and trail and public access improvements.
3. The Town of Esopus, is also working on a comprehensive plan that contains waterfront [60] goals for the Wallkill, Hudson, and Rondout Creek Waterfront access and usage. These plans include development policies to restore, revitalize, and redevelop deteriorated and underutilized waterfront areas for commercial, industrial, cultural, recreational, and other uses.
4. Significant work has also been done to address storms and sea level rise: Preparing Hudson River Waterfronts for an Era of Rapid Sea Level Rise [61], The City of Kingston Grant-Funded Waterfront Projects [62], Scenic Hudson's report on Sea Level Rise [63], and Hudson River Sea Level Rise City of Kingston [63]. New York is also starting to pay attention to the climate issue through The Climate Leadership and Community Protection Act (CLCPA) [64].

However, importantly, very few if any of these proposals are in the implementation phase. And little of the available climate and sea level change studies and data are included in the port and waterfront redevelopment plans as presently formulated.

Rondout Riverport 2040 is unique in that it takes likely forecasts of the near future fully into account; it is a proposal that offers a hard, sober look at the realities of our climate change, alternative energy, and global supply chain future.

But for this plan to be realized, stakeholders, partners, and existing maritime institutions will need to buy-in now and participate actively in the planning and implementation process. Those institutions include, but are not limited to the Center for Post Carbon Logistics [4], the Schooner Apollonia [2], the Hudson River Maritime Museum's solar electric passenger vessel Solaris, [1] the Hudson River Sloop Clearwater [65], Sustainable Hudson Valley's Regional Hudson Valley Climate Action Plan [66], The Riverport Coalition [67], Riverkeeper [68], and the Beacon Sloop Club's Woody Guthrie [69]. This diverse partnership must also be inclusive of public and private landowners, as well as land conservation organizations, including but not limited to the Kingston Land Trust [70], Scenic Hudson, and the hundreds of Hudson Valley organizations and individuals working for a more resilient and sustainable future.

4.1 Threat Assessment, a first step.

The Rondout communities must start by objectively assessing near future threats and evaluating our greatest points of weakness – assessing local infrastructure, and economic, political, social, and environmental structures. Rondout Riverport communities will especially need to fortify against the economic and environmental storms to come by doing work to enrich our towns and neighborhoods today, reducing risk and enhancing resilience for the future, by:

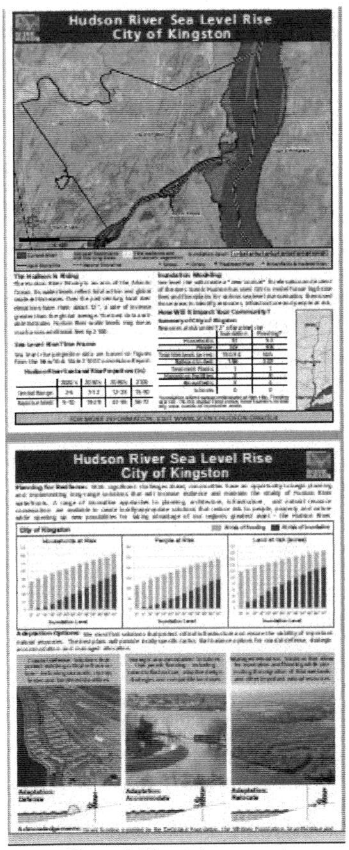

Figure 7. Threat Assessment

- Implementing community flood-proofing [71] and simultaneously introducing drought-resistant landscaping [72], in preparation for extreme weather patterns.

- institutionalizing green building practices [73];

- zoning against development in climate disaster-prone floodplains [74];

- installing redundant communication and storm-proof energy systems [75];

- establishing systems for community-wide food security [76];

- creating "resiliency hubs," [77] equipped to deal with sudden health, environmental, or weather / climate disasters, and develop strategies for proactive risk minimization and management;

- using the tools of community "placemaking" [78] to include the broadest possible participation in planning for, and developing, a working and recreational Rondout Waterfront and Port that will be operational and adaptable for the first half of the 21st century and beyond;

- develop an education and training center [36] for rapid proliferation of all these practices and trades in and beyond the Rondout Riverport.

5. PLACEMAKING - A PATHWAY TO THE FUTURE

Rondout Riverport 2040 will engender the Hudson Valley's can-do spirit, harness our region's inventiveness and our love of innovation, allowing our region and its people to not merely survive in the Post Carbon era, but thrive. And why not? After all, our region gave the world the steamboat, the telegraph, the submarine, FM radio, the first interactive software systems vital to today's computers, and even potato chips. We seem born to invent the future!

Rondout Riverport 2040, by cooperating fully with all partners, will incorporate the best elements of existing planning documents; undertake a thorough land use, flood plain, and sea level rise analysis; examine current trends in shipping, energy, food security and port management; assess the best climate change and economic forecasts; and create an adaptive re-use Waterfront plan that incorporates the best of 19th, 20th, and 21st century technologies.

But this process will do far more than construct a vision. It must ensure that this vision is aligned with community values and sensibilities. To achieve this goal, we will use a placemaking approach as the structure for addressing critical questions about how best to mobilize the many assets of the Rondout Riverport in a coordinated fashion to meet community needs and attract diverse resources.

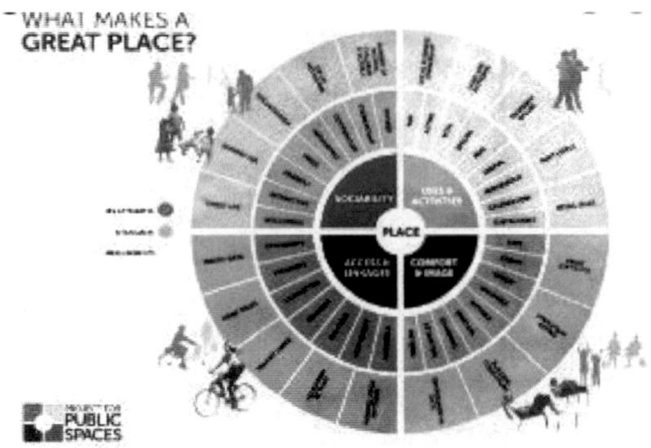

Figure 8. What Makes a Great Place?

Placemaking is a holistic approach for considering the possibilities inherent in a locality by identifying a unifying purpose or theme – the essence of the place – and then identifying multiple strategies, at multiple scales, that relate to this theme, providing direction for achieving unified objectives and goals.

The foundation of placemaking is a focus on the many natural benefits of public space to achieve the most comprehensive multiple uses, aesthetic benefits, connectivity, and social interaction. This process will generate key insights into how state, municipal, and county government agencies can best coordinate implementation efforts and find the resources to address problems and opportunities.

The placemaking approach will catalyze the integration of the many layers of conceptual planning already underway by various entities, aiding in the development of collaborative strategies for redeveloping the Port so that it serves multiple river uses and users.

The partners will work with, and gain consensus from, other Hudson Valley organizations to begin realizing the *Rondout Riverport 2040* vision. A network of groups, including the Lifeboats HV, Sustainable Hudson Valley Senior Fellows [79], Good Work Institute Fellows Network [80], C4PCL's advisory committee [81], plus staff and contractors, will provide intensive inputs and garner resources to translate the partners' vision into robust planning and implementation during 2021.

It is now past time to implement the many excellent ideas generated by our communities and their planners. It is time to bring the planning process forward into those communities. The path to a bright, sustainable future starts with research and engagement, and placemaking in Kingston and Esopus and on the Rondout Waterfront.

The source of our inspiration and empowerment will be our region's shorefront and its waters, its hands, and minds. Here the best and brightest, urban and rural, "Slow" technologists, craftspeople, educators, planners, artists, schoolchildren, and seniors, can come together to remake our post-modern world. Here we will find new, efficient, green ways to produce energy; revolutionize agriculture to assure food security; reinvent transportation on land and water to move goods up and down our Hudson and to prosper in the challenging times ahead. Here we'll help birth a new, inclusive regional economy that rewards all citizens, while celebrating democracy, cooperation, and public service.

Picture a Rondout Port in which every day, diverse participants – Transition and Permaculture practitioners, boat and ship builders, coopers, riggers, longshore workers, managers, carpenters, commercial fishermen, millwrights, engineers, potters, community development financial institutions, farmers, weavers, woodworkers, planners, architects, writers, historians, archivists, computer and IT experts, and people from wildly diverse vocations – will all merge and meld their talents to realize the vision of *Rondout Riverport 2040*.

In implementing the Rondout Riverport vision, we will move via hands-on experiences beyond spin and abstract buzzwords – past "environmental", or "sustainable", or "eco" this or that. Here, our work will focus on a single place and on a Just Transition away from fossil fuels. The times ahead will give new meaning to the word deckhand, as all join to create the naturally viable means for living and being in community in the 21st Century – as we prosper economically, emotionally, and spiritually, beyond the realm of coal and oil.

The next step will be one of the most critical: to gather all our research and data, analyze it, and commit to honestly confronting challenges, while also boldly embracing opportunities and possibilities. We must move forward quickly and vigorously; climate change and economic change are moving ahead swiftly. We must inspire individuals, communities, local leaders, and City, County, and State officials to commit to the creation of a thriving, innovative Rondout Riverport and Working Waterfront, as a gateway to a vast system of sustainable waterways that together will enable a Post Carbon Future full of hope and opportunity.

PHOTOGRAPHS AND CHARTS

[1] Deniella Zalcman for the Wall Street Journal
[2] Rope Walk https://www.tekportal.net/ropewalk/
[3] Manipulated photograph fromhttps://www.oxfordmail.co.uk/news/17709773.solar-shine-proof-narrow-boats-can-powered-sun/
[4] https://artgallery.yale.edu/collections/objects/47354
[5] https://www.scenichudson.org/our-work/climate/sea-level-rise/
[6] Andrew Willner manipulated photograph.
[7] https://www.slideshare.net/TeachingtheHudsonValley/investigating-sea-level-rise-in-the-hudson-valley
[8] https://www.playcore.com/successful-placemaking

REFERENCES

[1] 'Hudson River Maritime Museum', https://www.hrmm.org/, n.d.

[2] 'Schooner Apollonia', http://www.schoonerapollonia.com/, n.d.

[3] 'Sustainable Hudson Valley', https://sustainhv.org/, n.d.

[4] 'Center for Post Carbon Logistics', http://postcarbonlogistics.org, n.d.

[5] 'The 19th century solar engines of Augustin Mouchot, Abel Pifre, and John Ericsson', *Landartgenerator* https://landartgenerator.org/blagi/archives/2004, 29 Feb 2012

[6] 'The Clearwater Story', https://www.clearwater.org/about/the-clearwater-story/, n.d.

[7] 'EcoClipper', https://ecoclipper.org/, n.d.

[8] 'Jorne Langelaan - Salt Water On Your Lips', *Regenerative Agriculture Sector Accellerator* https://rasa.ag/jorne-langelaan-salt-water-on-your-lips/, 4 Oct 2020

[9] Freehill-Maye, Lynn, 'Carbon Neutral Shipping On The Hudson', *Scenic Hudson*, https://www.scenichudson.org/viewfinder/carbon-neutral-shipping-on-the-hudson/, n.d.

[10] Jones, Kevin, 'AutoBiography!', http://kkjones.net/bio.aspx n.d.

[11] 'Feeny Shipyard' https://feeneyshipyard.com/home n.d.

[12] 'Solar Sal' https://www.solarsal.solar/ n.d.

[13] 'Delaware and Hudson Canal and Gravity Railroad - Upper Delaware, NY/PA', *Scenic Wild Delaware River* https://scenicwilddelawareriver.com/entries/delaware-and-hudson-canal-and-gravity-railroad-upper-delaware-pany/3243c48f-039a-4f67-a50e-3f28bd73c883, n.d.

[14] 'Quercus Cooperage' https://www.qcooperage.com/ n.d.

[15] D. Suresh, Kishore, P.K. Rohatgi, 'Working experience with a foundry solar furnace', *Solar Energy*, Vol 27, Issue 6, 1981, Pages 457-462, http://www.sciencedirect.com/science/article/pii/0038092X81900414

[16] Floyd, Charlie, 'How rope is made at the only traditional working ropewalk in the world', *Insider* https://www.insider.com/how-traditional-rope-is-made-chatham-dockyard-ships-ropewalk-2018-11, 16 Nov 2018

[17] Park, Haeun, 'Rice paddies in Vermont? Vergennes farm boasts largest rice operation in Northeast' *The Middlebury Campus* https://middleburycampus.com/46445/local/rice-paddies-in-vermont-vergennes-farm-boasts-largest-rice-operation-in-northeast/, 3 Oct 2019.

[18] Andrus, Erik, 'Sailing like Stink' *The Vermont Sail Freight Project* https://vermontsailfreightproject.wordpress.com/2014/06/12/, 12 June 2014

[19] de Grailly, Victor, 'View near Anthony's Nose - Hudson River, NY' *Mutual Art* https://www.mutualart.com/Artwork/-View-Near-Anthony-s-Nose---Hudson-River/7F4D50E5BE594838 n.d.

[20] NYS Department of Environmental Conservation, 'Coastal Resiliency Actions' https://www.dec.ny.gov/lands/111216.html#Climate 17 July 2020

[21] Messner, S., Moran, L., Reub, G., & Campbell, J., 'Climate Change and sea level rise impacts at Ports and a consistent methodology to evaluate vulnerability and risk', *WIT Transactions on Ecology and the Environment* Vol 169, Oct 2013 https://www.researchgate.net/publication/257875273_Climate_change_and_sea_level_rise_impacts_at_Ports_and_a_consistent_methodology_to_evaluate_vulnerability_and_risk

[22] Scenic Hudson, 'Hudson River Sea Level Rise City Of Kingston' http://www.scenichudson.org/wp-content/uploads/legacy/SLRCityofKingston.pdf, n.d.

[23] Chen, James, 'Impact Investing Definition' *Investopedia* https://www.investopedia.com/terms/i/impact-investing.asp, 9 Oct 2019

[24] Gilbert, Richard, and Perl, Anthony, 'Transportation in the Post-Carbon World' *The Post Carbon Reader* https://www.resilience.org/stories/2010-10-27/transportation-transportation-post-carbon-world/, 2010

[25] Sun, Christina, and Craft, Carter, 'From Trucks to Tugs: Short Sea Shipping' *Urban Omnibus* https://urbanomnibus.net/2011/05/from-trucks-to-tugs-short-sea-shipping/, 25 May 2011

[26] New York State Canal Corporation, 'Canal Map - New York State Canals.' http://www.canals.ny.gov/maps/ n.d.

[27] 'Voices from the Working Waterfront' *NOAA* https://oceanservice.noaa.gov/news/jul14/working-waterfront.html n.d.

[28] Hsieh, Chia-wen Carmen, and Felby, Claus, 'Biofuels for the Marine Shipping Sector' IEA Bioenergy, 2017 http://task39.sites.olt.ubc.ca/files/2013/05/Marine-biofuel-report-final-Oct-2017.pdf

[29] Sahar, Sana Sadaf, Javed Iqbal, Inam Ullah, Haq Nawaz Bhatti, Shazia Nouren, Habib-ur-Rehman, Jan Nisar, Munawar Iqbal, 'Biodiesel production from waste cooking oil: An efficient technique to convert waste into biodiesel' *Sustainable Cities and Society* Vol 41, 2018 Pages 220-226, https://www.sciencedirect.com/science/article/abs/pii/S2210670717317249

[30] Weston, Pheoby, and Yagoda, Maya, 'World's first Hydrogen-Powered ship docks in London as

part of Zero-Emissions global tour.' *Independent* https://www.independent.co.uk/environment/energy-observer-hydrogen-powered-zero-emissions-ship-london-climate-a9146571.html 7 Oct 2019
[31] DEAD LINK https://www.fema.gov/media-library-data/1404150030143-cd3760624f61032d097df173e7f18355/FEMA_P312_Chap_7.pdf
[32] 'Rondout Woodworking' http://www.rondoutwoodworking.com/ n.d.
[33] 'What is Break Bulk?' https://www.crowley.com/news-and-media/blog/what-is-breakbulk/ n.d.
[34] 'Types of packaging and stowage methods for break bulk cargo' *General Cargo Ship* http://generalcargoship.com/types-of-packaging.html n.d.
[35] 35. Hu, Winnie, and Haag, Matthew, 'Park it, Trucks: Here come New York's cargo bikes', *The New York Times* https://www.nytimes.com/2019/12/04/nyregion/nyc-cargo-bikes-delivery.html 4 Dec 2019
[36] 36. Willner, Andrew, 'Wellbeing farm, a slow tech living laboratory for the Hudson Valley' *Center for Post Carbon Logistics* https://postcarbonlogistics.org/2020/08/25/wellbeing-farm-a-slow-tech-living-laboratory-for-the-hudson-valley-bioregion/ 25 Aug 2020
[37] 'Transition US' https://transitionus.org/ n.d.
[38] Doyon, Marie, 'A Geothermal Greenhouse Grow Microgreens All Year' *Upstate House* https://upstatehouse.com/geothermal-greenhouse-grows-microgreens-year/, 2017
[39] 'Hunt's Point Lifelines' *Rebuild by Design* http://www.rebuildbydesign.org/our-work/all-proposals/winning-projects/hunts-point-lifelines, n.d.
[40] 'Wooden Boat School' https://www.hrmm.org/boat-school.html n.d.
[41] 'New York Harbor School' https://www.newyorkharborschool.org/ n.d.
[42] 'The economic contribution of working waterfronts' *NOAA* https://coast.noaa.gov/digitalcoast/training/working-waterfronts.html n.d.
[43] 'Benefits of Bulkhead Removal' *Northwest Straits Foundation* https://nwstraitsfoundation.org/project/shoreline-armor-reduction-program-sharp-7/ n.d.
[44] 'Permaculture Design Principles' *Permaculture Principals* https://permacultureprinciples.com/principles/ n.d.
[45] Willner, Andrew, 'Transition, Permaculture, and Slow Technology' *Center for Post Carbon Logistics* https://postcarbonlogistics.org/2019/10/20/transition-permaculture-and-slow-technology/ 20 Oct 2019
[46] US Department of Transportation, 'Development of short sea shipping' https://www.transportation.gov/testimony/development-short-sea-shipping 15 Feb 2007
[47] Melotti, Rob, 'Trading Under Sail' *Practical Boat Owner* https://static1.squarespace.com/static/579681038419c282935ae899/t/5ac791280e2e72e497bb90f6/1523028269102/PBO+Sail+Cargo.pdf n.d.
[48] US Department of Transport 'America's Marine Highway' https://www.maritime.dot.gov/grants/marine-highways/marine-highway n.d.
[49] 'Neoline' https://www.neoline.eu/en/ n.d.
[50] Brownski 'Rondout Creek: Paddle back in time' *NY Ski Blog* https://nyskiblog.com/rondout-creek-kayaking-paddle-in-time/ 9 July 2018
[51] Finkle, Steven, 'Waterfront Brownfield Opportunity Area' https://www.kingston-ny.gov/filestorage/8463/8511/8682/8690/KingstonBOA-Step2-Narrative-April2011.pdf April 2011
[52] 'Combined Sewer Overflow Information' https://kingston-ny.gov/cso n.d.
[53] New York Department of State 'Coastal Policies' https://www.dos.ny.gov/opd/programs/pdfs/CoastalPolicies.pdf 2017
[54] Kirby, Paul, 'Developer proposes a 4-acre, multiuse complex at Rondout waterfront' *Daily Freeman* https://www.dailyfreeman.com/news/local-news/developer-proposes--acre-multiuse-complex-at-rondout-waterfront/article_786d759c-e9c5-11e8-84eb-3372fc2b0561.html 17 Nov 2017
[55] 'Weaving the Waterfront' https://www.kingston-ny.gov/weavingthewaterfront n.d.
[56] 'Weaving the Waterfront Kingston' https://kingston-ny.gov/filestorage/8399/8469/26768/12450/31097/191023_Kingtson_Shoreline-FINAL.pdf n.d.
[57] 'Kingston Point Park Improvements' https://kingston-ny.gov/filestorage/8399/8469/26768/12450/31097/191022_Kingston_Point_Park-FINAL.pdf n.d.
[58] 'Kingston Point Climate Adaptive Design' https://kingston-ny.gov/filestorage/8399/8469/26768/12450/31097/191022_Kingston_Point_CAD-FINAL_rev1.pdf n.d.
[59] 'Kingston Point Rail Trail Phase 2' https://kingston-ny.gov/kprt2 n.d.
[60] Kemble, William J., 'Town of Esopus ready to move ahead with changes to waterfront plan.' *Daily Freeman* https://www.dailyfreeman.com/news/local-news/town-of-esopus-ready-to-move-ahead-with-changes-to-waterfront-plan/article_df9973f4-af23-11ea-8df7-8710a8c323bc.html 15 June 2020
[61] Swanzey, Gregg, and Tabak, Nava 'Preparing Hudson River waterfronts for an era of rapid sea level rise'

https://law.pace.edu/sites/default/files/LULC/Conference_2015/Swanzey%20Tabak%20PPT.pdf 2013

[62] 'City of Kingston Grant-Funded Waterfront Projects' https://www.kingston-ny.gov/filestorage/8399/8469/26768/12450/201804_City-funded_Waterfront_Projects_for_the_Rondout_Riverport_Community_Coalition_FINAL.pdf April 2018

[63] 'Sea Level Rise' https://www.scenichudson.org/our-work/climate/sea-level-rise/ n.d.

[64] 'Climate Act' https://climate.ny.gov/ n.d.

[65] 'The Sloop' https://www.clearwater.org/the-sloop/ n.d.

[66] 'Regional Climate Action Planning' *Sustainable Hudson Valley* https://sustainhv.org/programs/climate-solutions-lab/regional-climate-action-planning/ n.d.

[67] 'Waterfront Coalition unites Kingston Stakeholders' *Mid Hudson News* https://midhudsonnews.com/2018/04/13/waterfront-coalition-unites-kingston-stakeholders/ 13 April 2018

[68] 'Riverkeeper' https://www.riverkeeper.org/ n.d.

[69] 'The woody Guthrie Sloop' https://www.woodyguthrie.org/merchandise/wgsloop.htm n.d.

[70] 'Kingston Land Trust' https://kingstonlandtrust.org/ n.d.

[71] 'Floodproof your community' http://www.brighterfuturestogether.co.uk/brighter-futures-together-toolkit/flood-proof-your-community/ n.d.

[72] 'CityLab' *Bloomburg* https://www.bloomberg.com/citylab n.d.

[73] 'Green Building' https://archive.epa.gov/greenbuilding/web/html/ n.d.

[74] 'Floodplain Zoning' https://www.uwsp.edu/cnr-ap/clue/Documents/Water/Floodplain_Zoning_Chapter.pdf n.d.

[75] 'Hurricane Proof Energy Resiliency' *North American Clean Energy* https://www.nacleanenergy.com/articles/30913/hurricane-proof-energy-resiliency 3 May 2018

[76] 'What is Community Food Security?' *Food Security* http://foodsecurity.org/views_cfs_faq/ n.d.

[77] 'Resilience Hubs' *Urban Sustainability Director's Network* http://resilience-hub.org/ n.d.

[78] 'What is Placemaking?' *Project for Public Spaces* https://www.pps.org/article/what-is-placemaking n.d.

[79] 'Senior Fellows' *Sustainable Hudson Valley* https://sustainhv.org/about/our-team/senior-fellows/ n.d.

[80] 'The Fellowship' *The Good Work Institute* https://goodworkinstitute.org/the-fellowship/ n.d.

[81] 'Advisory Committee' *Center for Post Carbon Logistics* https://postcarbonlogistics.org/advisory-committee/ n.d.

AUTHOR BIOGRAPHY

Andrew Willner is the Executive Director of the Center for Post Carbon Logistics, and from 2008 to 2014 was the Principal Professional Consultant for Energy, Transportation, and the Environment to the Hugo Neu Corporation He has been a leader, organizer, and advocate for the New York/New Jersey Bioregion for almost 40 years. He was an early proponent of the Waterkeeper model of water and habitat protection as the founder of NY/NJ Baykeeper. He has experience as a city planner, wood worker, boat builder, photographer, Permaculture and Transition practitioner, storyteller, and author. Also, he is a sought-after speaker on a wide variety of subjects including post carbon logistics, environmental advocacy, habitat restoration, and regenerative communities.